THE
DISAPPEARING
SPOON

THE DISAPPEARING SPOON

And Other True Tales of Madness,
Love, and the History of the World from
the Periodic Table of the Elements

S AM   K EAN

Doubleday

LONDON · TORONTO · SYDNEY · AUCKLAND · JOHANNESBURG

TRANSWORLD PUBLISHERS
61–63 Uxbridge Road, London W5 5SA
A Random House Group Company
www.rbooks.co.uk

First published in the United States of America
in 2010 by Little, Brown and Company
a division of Hachette Book Group, Inc.

First published in Great Britain
in 2011 by Doubleday
an imprint of Transworld Publishers

A CIP catalogue record for this book
is available from the British Library.

ISBNs 9780857520265 (cased)
9780857520272 (tpb)

Addresses for Random House Group Ltd companies outside the UK
can be found at: www.randomhouse.co.uk
The Random House Group Ltd Reg. No. 954009

The Random House Group Ltd supports the Forest Stewardship Council (FSC),
the leading international forest-certification organization. All our titles that are
printed on Greenpeace-approved FSC-certified paper carry the FSC logo.
Our paper procurement policy can be found at
www.rbooks.co.uk/environment

Typeset in Janson
Printed and bound in Great Britain by
Clays Ltd, Bungay, Suffolk

2 4 6 8 10 9 7 5 3 1

Mixed Sources
Product group from well-managed
forests and other controlled sources
www.fsc.org Cert no. TT-COC-2139
© 1996 Forest Stewardship Council
FSC

CONTENTS

Contents

THE
DISAPPEARING
SPOON

INTRODUCTION

As a child in the early 1980s, I tended to talk with things in my mouth—food, dentist's tubes, balloons that would fly away, whatever—and if no one else was around, I'd talk anyway. This habit led to my fascination with the periodic table the first time I was left alone with a thermometer under my tongue. I came down with strep throat something like a dozen times in the second and third grades, and for days on end it would hurt to swallow. I didn't mind staying home from school and medicating myself with vanilla ice cream and chocolate sauce. Being sick always gave me another chance to break an old-fashioned mercury thermometer, too.

Lying there with the glass stick under my tongue, I would answer an imagined question out loud, and the thermometer would slip from my mouth and shatter on the hardwood floor, the liquid mercury in the bulb scattering like ball bearings. A minute later, my mother would drop to the floor despite her arthritic hip and begin corralling the balls. Using a toothpick like a hockey stick, she'd brush the supple spheres toward one another until they almost touched. Suddenly, with a final nudge, one sphere would gulp the other. A single, seamless ball would be left quivering where there had been two. She'd repeat this magic trick over and over across the floor, one large ball swallowing the others until the entire silver lentil was reconstructed.

Once she'd gathered every bit of mercury, she'd take down the green-labeled plastic pill bottle that we kept on a knick-knack shelf in the kitchen between a teddy bear with a fishing pole and a blue ceramic mug from a 1985 family reunion. After rolling the ball onto an envelope, she'd carefully pour the latest thermometer's worth of mercury onto the pecan-sized glob in the bottle. Sometimes, before hiding the bottle away, she'd pour the quicksilver into the lid and let my siblings and me watch the futuristic metal whisk around, always splitting and healing itself flawlessly. I felt pangs for children whose mothers so feared mercury they wouldn't even let them eat tuna. Medieval alchemists, despite their lust for gold, considered mercury the most potent and poetic substance in the universe. As a child I would have agreed with them. I would even have believed, as they did, that it transcended pedestrian categories of liquid or solid, metal or water, heaven or hell; that it housed otherworldly spirits.

Mercury acts this way, I later found out, because it is an element. Unlike water (H_2O), or carbon dioxide (CO_2), or almost anything else you encounter day to day, you cannot naturally separate mercury into smaller units. In fact, mercury is one of the more cultish elements: its atoms want to keep company only with other mercury atoms, and they minimize contact with the outside world by crouching into a sphere. Most liquids I spilled as a child weren't like that. Water tumbled all over, as did oil, vinegar, and unset Jell-O. Mercury never left a speck. My parents always warned me to wear shoes whenever I dropped a thermometer, to prevent those invisible glass shards from getting into my feet. But I never recall warnings about stray mercury.

For a long time, I kept an eye out for element eighty at school and in books, as you might watch for a childhood friend's name in the newspaper. I'm from the Great Plains and had learned in history class that Lewis and Clark had trekked through South Dakota and the rest of the Louisiana Territory with a microscope, compasses, sextants, three mercury thermometers, and other instruments. What I didn't know at first is that they also carried with them six hundred mercury laxatives, each four times the size of an aspirin. The laxatives were called Dr. Rush's Bilious Pills, after Benjamin Rush, a signer of the Declaration of Independence and a medical hero for bravely staying in Philadelphia during a yellow fever epidemic in 1793. His pet treatment, for any disease, was a mercury-chloride sludge administered orally. Despite the progress medicine made overall between 1400 and 1800, doctors in that era remained closer to medicine men than medical men. With a sort of sympathetic magic, they figured that beautiful, alluring mercury could cure patients by bringing them to an ugly crisis—poison fighting poison. Dr. Rush made patients ingest the solution until they drooled, and often people's teeth and hair fell out after weeks or months of continuous treatment. His "cure" no doubt poisoned or outright killed swaths of people whom yellow fever might have spared. Even so, having perfected his treatment in Philadelphia, ten years later he sent Meriwether and William off with some prepackaged samples. As a handy side effect, Dr. Rush's pills have enabled modern archaeologists to track down campsites used by the explorers. With the weird food and questionable water they encountered in the wild, someone in their party was always queasy, and to this day, mercury deposits dot the soil many places where the gang dug a latrine, perhaps after one of Dr. Rush's "Thunderclappers" had worked a little too well.

Mercury also came up in science class. When first presented with the jumble of the periodic table, I scanned for mercury and couldn't find it. It is there—between gold, which is also dense and soft, and thallium, which is also poisonous. But the symbol for mercury, Hg, consists of two letters that don't even appear in its name. Unraveling that mystery—it's from *hydragyrum*, Latin for "water silver"—helped me understand how heavily ancient languages and mythology influenced the periodic table, something you can still see in the Latin names for the newer, superheavy elements along the bottom row.

I found mercury in literature class, too. Hat manufacturers once used a bright orange mercury wash to separate fur from pelts, and the common hatters who dredged around in the steamy vats, like the mad one in *Alice in Wonderland*, gradually lost their hair and wits. Eventually, I realized how poisonous mercury is. That explained why Dr. Rush's Bilious Pills purged the bowels so well: the body will rid itself of any poison, mercury included. And as toxic as swallowing mercury is, its fumes are worse. They fray the "wires" in the central nervous system and burn holes in the brain, much as advanced Alzheimer's disease does.

But the more I learned about the dangers of mercury, the more—like William Blake's "Tyger! Tyger! burning bright"—its destructive beauty attracted me. Over the years, my parents redecorated their kitchen and took down the shelf with the mug and teddy bear, but they kept the knickknacks together in a cardboard box. On a recent visit, I dug out the green-labeled bottle and opened it. Tilting it back and forth, I could feel the weight inside sliding in a circle. When I peeked over the rim, my eyes fixed on the tiny bits that had splashed to the sides of the main channel. They just sat there, glistening, like beads of water so perfect you'd encounter them only in fantasies. All throughout my childhood, I associated spilled mercury with a fever. This time, knowing the fearful symmetry of those little spheres, I felt a chill.

* * *

From that one element, I learned history, etymology, alchemy, mythology, literature, poison forensics, and psychology.* And those weren't the only elemental stories I collected, especially after I immersed myself in scientific studies in college and found a few professors who gladly set aside their research for a little science chitchat.

As a physics major with hopes of escaping the lab to write, I felt miserable among the serious and gifted young scientists in my classes, who loved trial-and-error experiments in a way I never could. I stuck out five frigid years in Minnesota and ended up with an honors degree in physics, but despite spending hundreds of hours in labs, despite memorizing thousands of equations, despite drawing tens of thousands of diagrams with frictionless pulleys and ramps—my real education was in my professors' stories. Stories about Gandhi and Godzilla and a eugenicist who used germanium to steal a Nobel Prize. About throwing blocks of explosive sodium into rivers and killing fish. About people suffocating, quite blissfully, on nitrogen gas in space shuttles. About a former professor on my campus who would experiment on the plutonium-powered pacemaker *inside his own chest*, speeding it up and slowing it down by standing next to and fiddling with giant magnetic coils.

I latched on to those tales, and recently, while reminiscing about mercury over breakfast, I realized that there's a funny, or odd, or chilling tale attached to every element on the periodic table. At the same time, the table is one of the great intellectual achievements of humankind. It's both a scientific accomplishment and a storybook, and I wrote this book to peel back all of

---

*This and all upcoming asterisks refer to the Notes and Errata section, which begins on page 349 and continues the discussion of various interesting points. Also, if you need to refer to a periodic table, see page 392.

its layers one by one, like the transparencies in an anatomy text-book that tell the same story at different depths. At its simplest level, the periodic table catalogs all the different kinds of matter in our universe, the hundred-odd characters whose head-strong personalities give rise to everything we see and touch. The shape of the table also gives us scientific clues as to how those personalities mingle with one another in crowds. On a slightly more complicated level, the periodic table encodes all sorts of forensic information about where every kind of atom came from and which atoms can fragment or mutate into different atoms. These atoms also naturally combine into dynamic systems like living creatures, and the periodic table predicts how. It even predicts what corridors of nefarious elements can hobble or destroy living things.

The periodic table is, finally, an anthropological marvel, a human artifact that reflects all of the wonderful and artful and ugly aspects of human beings and how we interact with the physical world—the history of our species written in a compact and elegant script. It deserves study on each of these levels, starting with the most elementary and moving gradually upward in complexity. And beyond just entertaining us, the tales of the periodic table provide a way of understanding it that never appears in textbooks or lab manuals. We eat and breathe the periodic table; people bet and lose huge sums on it; philosophers use it to probe the meaning of science; it poisons people; it spawns wars. Between hydrogen at the top left and the man-made impossibilities lurking along the bottom, you can find bubbles, bombs, money, alchemy, petty politics, history, poison, crime, and love. Even some science.

Part I

ORIENTATION: COLUMN BY COLUMN, ROW BY ROW

1

Geography Is Destiny

When most people think of the periodic table, they
remember a chart hanging on the front wall of their high
school chemistry class, an asymmetric expanse of columns and
rows looming over one of the teacher's shoulders. The chart was
usually enormous, six by four feet or so, a size both daunting
and appropriate, given its importance to chemistry. It was introduced to the class in early September and was still relevant in
late May, and it was the one piece of scientific information that,
unlike lecture notes or textbooks, you were encouraged to con-
sult during exams. Of course, part of the frustration you might
remember about the periodic table could flow from the fact
that, despite its being freely available to fall back on, a gigantic
and fully sanctioned cheat sheet, it remained less than frickin'
helpful.

On the one hand, the periodic table seemed organized and
honed, almost German engineered for maximum scientific utility. On the other hand, it was such a jumble of long numbers,
abbreviations, and what looked for all the world like computer
error messages ([Xe]$6s^2 4f^1 5d^1$), it was hard not to feel anxious.
And although the periodic table obviously had something to do
with other sciences, such as biology and physics, it wasn't clear

what exactly. Probably the biggest frustration for many students was that the people who *got* the periodic table, who could really unpack how it worked, could pull so many facts from it with such dweeby nonchalance. It was the same irritation color-blind people must feel when the fully sighted find sevens and nines lurking inside those parti-colored dot diagrams—crucial but hidden information that never quite resolves itself into coherence. People remember the table with a mix of fascination, fondness, inadequacy, and loathing.

Before introducing the periodic table, every teacher should strip away all the clutter and have students just stare at the thing, blank.

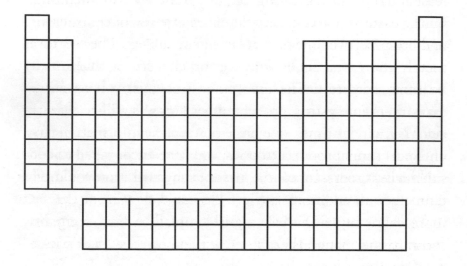

What does it look like? Sort of like a castle, with an uneven main wall, as if the royal masons hadn't quite finished building up the left-hand side, and tall, defensive turrets on both ends. It has eighteen jagged columns and seven horizontal rows, with a "landing strip" of two extra rows hanging below. The castle is made of "bricks," and the first non-obvious thing about it is that the bricks are not interchangeable. Each brick is an *element*, or type of substance (as of now, 112 elements, with a few more pending, make up the table), and the entire castle would crumble if any of those bricks didn't sit exactly where it does. That's no exaggeration: if scientists determined that one element somehow fit into a different slot or that two of the elements could be swapped, the entire edifice would tumble down.

Another architectural curiosity is that the castle is made up of different materials in different areas. That is, not all the bricks are made of the same substance, nor do they have the same characteristics. Seventy-five percent of the bricks are metals, which means most elements are cold, gray solids, at least at temperatures human beings are used to. A few columns on the eastern side contain gases. Only two elements, mercury and bromine, are liquids at room temperature. In between the metals and gases, about where Kentucky sits on a U.S. map, lie some hard-to-define elements, whose amorphous nature gives them interesting properties, such as the ability to make acids billions of times stronger than anything locked up in a chemical supply room. Overall, if each brick was made of the substance it represented, the castle of the elements would be a chimera with additions and wings from incongruent eras, or, more charitably, a Daniel Libeskind building, with seemingly incompatible materials grafted together into an elegant whole.

The reason for lingering over the blueprints of the castle walls is that the coordinates of an element determine nearly everything scientifically interesting about it. For each element, its geography is its destiny. In fact, now that you have a sense of what the table looks like in outline, I can switch to a more useful metaphor: the periodic table as a map. And to sketch in a bit more detail, I'm going to plot this map from east to west, lingering over both well-known and out-of-the-way elements.

First up, in column eighteen at the far right-hand side, is a set of elements known as the noble gases. *Noble* is an archaic, funny-sounding word, less chemistry than ethics or philosophy. And indeed, the term "noble gases" goes back to the birthplace of Western philosophy, ancient Greece. There, after his fellow Greeks Leucippus and Democritus invented the idea of atoms, Plato minted the word "elements" (in Greek, *stoicheia*) as a general term for different small particles of matter. Plato—who left Athens for his own safety after the death of his mentor, Socrates, around 400 BC and wandered around writing philosophy for years—of course lacked knowledge of what an element really is in chemistry terms. But if he had known, he no doubt would have selected the elements on the eastern edge of the table, especially helium, as his favorites.

In his dialogue on love and the erotic, *The Symposium*, Plato claimed that every being longs to find its complement, its missing half. When applied to people, this implies passion and sex and all the troubles that accompany passion and sex. In addition, Plato emphasized throughout his dialogues that abstract and unchanging things are intrinsically more noble than things that grub around and interact with gross matter. This explains why he adored geometry, with its idealized circles and cubes, objects perceptible only to our reason. For nonmathematical objects, Plato developed a theory of "forms," which argued that all objects are shadows of one ideal type. All trees, for instance, are imperfect copies of an ideal tree, whose perfect "tree-ness" they aspire to. The same with fish and "fish-ness" or even cups and "cup-ness." Plato believed that these forms were not merely theoretical but actually existed, even if they floated around in an empyrean realm beyond the direct perception of humans. He would have been as shocked as anyone, then, when scientists began conjuring up ideal forms on earth with helium.

In 1911, a Dutch-German scientist was cooling mercury with liquid helium when he discovered that below –452°F the system lost all electrical resistance and became an ideal conductor. This would be sort of like cooling an iPod down to hundreds of degrees below zero and finding that the battery remained fully charged no matter how long or loud you played music, until infinity, as long as the helium kept the circuitry cold. A Russian-Canadian team pulled an even neater trick in 1937 with pure helium. When cooled down to –456°F, helium turned into a superfluid, with exactly zero viscosity and zero resistance to flow—perfect fluidness. Superfluid helium defies gravity and flows uphill and over walls. At the time, these were flabbergasting finds. Scientists often fudge and pretend that effects like friction equal zero, but only to simplify calculations. Not even Plato predicted someone would actually find one of his ideal forms.

Helium is also the best example of "element-ness"—a substance that cannot be broken down or altered by normal, chemical means. It took scientists 2,200 years, from Greece in 400 BC to Europe in 1800 AD, to grasp what elements really are, because most are too changeable. It was hard to see what made carbon *carbon* when it appeared in thousands of compounds, all with different properties. Today we would say that carbon dioxide, for instance, isn't an element because one molecule of it divides into carbon and oxygen. But carbon and oxygen *are* elements because you cannot divide them more finely without destroying them. Returning to the theme of *The Symposium* and Plato's theory of erotic longing for a missing half, we find that virtually every element seeks out other atoms to form bonds with, bonds that mask its nature. Even most "pure" elements, such as oxygen molecules in the air (O_2), always appear as composites in nature. Yet scientists might have figured out what elements are much sooner had they known about helium, which has never reacted with another substance, has never been anything but a pure element.*

Helium acts this way for a reason. All atoms contain negative particles called electrons, which reside in different tiers, or energy levels, inside the atom. The levels are nested concentrically inside each other, and each level needs a certain number of electrons to fill itself and feel satisfied. In the innermost level, that number is two. In other levels, it's usually eight. Elements normally have equal numbers of negative electrons and positive particles called protons, so they're electrically neutral. Electrons, however, can be freely traded between atoms, and when atoms lose or gain electrons, they form charged atoms called ions.

What's important to know is that atoms fill their inner, lower-energy levels as full as possible with their own electrons, then either shed, share, or steal electrons to secure the right number in the outermost level. Some elements share or trade electrons diplomatically, while others act very, very nasty. That's half of chemistry in one sentence: atoms that don't have enough electrons in the outer level will fight, barter, beg, make and break alliances, or do whatever they must to get the right number.

Helium, element two, has exactly the number of electrons it needs to fill its only level. This "closed" configuration gives helium tremendous independence, because it doesn't need to interact with other atoms or share or steal electrons to feel satisfied. Helium has found its erotic complement in itself. What's more, that same configuration extends down the entire eighteenth column beneath helium—the gases neon, argon, krypton, xenon, and radon. All these elements have closed shells with full complements of electrons, so none of them reacts with anything under normal conditions. That's why, despite all the fervid activity to identify and label elements in the 1800s—including the development of the periodic table itself—no one isolated a single gas from column eighteen before 1895. That aloofness from everyday experience, so like his ideal spheres and triangles, would have charmed Plato. And it was that sense the scientists who discovered helium and its brethren on earth were trying to evoke with the name "noble gases." Or to put it in Plato-like words, "He who adores the perfect and unchangeable and scorns the corruptible and ignoble will prefer the noble gases, by far, to all other elements. For they never vary, never waver, never pander to other elements like hoi polloi offering cheap wares in the marketplace. They are incorruptible and ideal."

The repose of the noble gases is rare, however. One column to the west sits the most energetic and reactive gases on the periodic table, the halogens. And if you think of the table wrapping around like a Mercator map, so that east meets west and column eighteen meets column one, even more violent elements appear on the western edge, the alkali metals. The pacifist noble gases are a demilitarized zone surrounded by unstable neighbors.

Despite being normal metals in some ways, the alkalis, instead of rusting or corroding, can spontaneously combust in air or water. They also form an alliance of interests with the halogen gases. The halogens have seven electrons in the outer layer, one short of the octet they need, while the alkalis have one electron in the outer level and a full octet in the level below. So it's natural for the latter to dump their extra electron on the former and for the resulting positive and negative ions to form strong links.

This sort of linking happens all the time, and for this reason electrons are the most important part of an atom. They take up virtually all an atom's space, like clouds swirling around an atom's compact core, the nucleus. That's true even though the components of the nucleus, protons and neutrons, are far bigger than individual electrons. If an atom were blown up to the size of a sports stadium, the proton-rich nucleus would be a tennis ball at the fifty-yard line. Electrons would be pinheads flashing around it—but flying so fast and knocking into you so many times per second that you wouldn't be able to enter the stadium: they'd feel like a solid wall. As a result, whenever atoms touch, the buried nucleus is mute; only the electrons matter.*

One quick caveat: Don't get too attached to the image of electrons as discrete pinheads flashing about a solid core. Or, in the more usual metaphor, don't necessarily think of electrons as planets circling a nucleic sun. The planet analogy is useful, but as with any analogy, it's easy to take too far, as some renowned scientists have found out to their chagrin.

Bonding between ions explains why combinations of halogens and alkalis, such as sodium chloride (table salt), are common. Similarly, elements from columns with two extra electrons, such as calcium, and elements from columns that need two extra electrons, such as oxygen, frequently align themselves. It's the easiest way to meet everyone's needs. Elements from nonreciprocal columns also match up according to the same laws. Two ions of sodium (Na^+) take on one of oxygen (O^{-2}) to form sodium oxide, Na_2O. Calcium chloride combines as $CaCl_2$ for the same reasons. Overall, you can usually tell at a glance how elements will combine by noting their column numbers and figuring out their charges. The pattern all falls out of the table's pleasing left-right symmetry.

Unfortunately, not all of the periodic table is so clean and neat. But the raggedness of some elements actually makes them interesting places to visit.

*   *   *

There's an old joke about a lab assistant who bursts into a scientist's office one morning, hysterical with joy despite a night of uninterrupted work. The assistant holds up a fizzing, hissing, corked bottle of green liquid and exclaims he has discovered a universal solvent. His sanguine boss peers at the bottle and asks, "And what is a universal solvent?" The assistant sputters, "An acid that dissolves all substances!"

After considering this thrilling news—not only would this universal acid be a scientific miracle, it would make both men billionaires—the scientist replies, "How are you holding it in a glass bottle?"

It's a good punch line, and it's easy to imagine Gilbert Lewis smiling, perhaps poignantly. Electrons drive the periodic table, and no one did more than Lewis to elucidate how electrons behave and form bonds in atoms. His electron work was especially illuminating for acids and bases, so he would have appreciated the assistant's absurd claim. More personally, the punch line might have reminded Lewis how fickle scientific glory can be.

A wanderer, Lewis grew up in Nebraska, attended college and graduate school in Massachusetts around 1900, and then studied in Germany under chemist Walther Nernst. Life under Nernst proved so miserable, for legitimate and merely perceived reasons, that Lewis returned to Massachusetts for an academic post after a few months. That, too, proved unhappy, so he fled to the newly conquered Philippines to work for the U.S. government, taking with him only one book, Nernst's *Theoretical Chemistry*, so he could spend years rooting out and obsessively publishing papers on every quibbling error.*

Eventually, Lewis grew homesick and settled at the University of California at Berkeley, where, over forty years, he built

Berkeley's chemistry department into the world's best. Though that may sound like a happy ending, it wasn't. The singular fact about Lewis is that he was probably the best scientist never to win the Nobel Prize, and he knew it. No one ever received more nominations, but his naked ambition and a trail of disputes worldwide poisoned his chances of getting enough votes. He soon began resigning (or was forced to resign) from prestigious posts in protest and became a bitter hermit.

Apart from personal reasons, Lewis never secured the Nobel Prize because his work was broad rather than deep. He never discovered one amazing thing, something you could point to and say, Wow! Instead, he spent his life refining how an atom's electrons work in many contexts, especially the class of molecules known as acids and bases. In general, whenever atoms swap electrons to break or form new bonds, chemists say they've "reacted." Acid-base reactions offer a stark and often violent example of those swaps, and Lewis's work on acids and bases did as much as anyone's to show what exchanging electrons means on a submicroscopic level.

Before about 1890, scientists judged acids and bases by tasting or dunking their fingers in them, not exactly the safest or most reliable methods. Within a few decades, scientists realized that acids were in essence proton donors. Many acids contain hydrogen, a simple element that consists of one electron circling one proton (that's all hydrogen has for a nucleus). When an acid like hydrochloric acid (HCl) mixes with water, it fissures into H^+ and Cl^-. Removing the negative electron from the hydrogen leaves just a bare proton, the H^+, which swims away on its own. Weak acids like vinegar pop a few protons into solution, while strong acids like sulfuric acid flood solutions with them.

Lewis decided this definition of an acid limited scientists

too much, since some substances act like acids without relying on hydrogen. So Lewis shifted the paradigm. Instead of saying that H^+ splits off, he emphasized that Cl^- absconds with its electron. Instead of a proton donor, then, an acid is an electron thief. In contrast, bases such as bleach or lye, which are the opposites of acids, might be called electron donors. These definitions, in addition to being more general, emphasize the behavior of electrons, which fits better with the electron-dependent chemistry of the periodic table.

Although Lewis laid this theory out in the 1920s and 1930s, scientists are still pushing the edge of how strong they can make acids using his ideas. Acid strength is measured by the pH scale, with lower numbers being stronger, and in 2005 a chemist from New Zealand invented a boron-based acid called a carborane, with a pH of −18. To put that in perspective, water has a pH of 7, and the concentrated HCl in our stomachs has a pH of 1. But according to the pH scale's unusual accounting methods, dropping one unit (e.g., from 4 to 3) boosts an acid's strength by ten times. So moving from stomach acid, at 1, to the boron-based acid, at −18, means the latter is ten billion billion times stronger. That's roughly the number of atoms it would take to stack them to the moon.

There are even worse acids based on antimony, an element with probably the most colorful history on the periodic table.* Nebuchadnezzar, the king who built the Hanging Gardens of Babylon in the sixth century BC, used a noxious antimony-lead mix to paint his palace walls yellow. Perhaps not coincidentally, he soon went mad, sleeping outdoors in fields and eating grass like an ox. Around that same time, Egyptian women were applying a different form of antimony as mascara, both to decorate their faces and to give themselves witchlike powers to cast the evil eye on enemies. Later, medieval monks—not to mention Isaac Newton—grew obsessed with the sexual properties of antimony and decided this half metal, half insulator, neither one thing nor the other, was a hermaphrodite. Antimony pills also won fame as laxatives. Unlike modern pills, these hard antimony pills didn't dissolve in the intestines, and the pills were considered so valuable that people rooted through fecal matter to retrieve and reuse them. Some lucky families even passed down laxatives from father to son. Perhaps for this reason, antimony found heavy work as a medicine, although it's actually toxic. Mozart probably died from taking too much to combat a severe fever.

Scientists eventually got a better handle on antimony. By the 1970s, they realized that its ability to hoard electron-greedy elements around itself made it wonderful for building custom acids. The results were as astounding as the helium superfluids. Mixing antimony pentafluoride, SbF_5, with hydrofluoric acid, HF, produces a substance with a pH of -31. This superacid is 100,000 billion billion billion times more potent than stomach acid and will eat through glass, as ruthlessly as water through paper. You couldn't pick up a bottle of it because after it ate through the bottle, it would dissolve your hand. To answer the professor in the joke, it's stored in special Teflon-lined containers.

To be honest, though, calling the antimony mix the world's strongest acid is kind of cheating. By themselves, SbF_5 (an electron thief) and HF (a proton donor) are nasty enough. But you have to sort of multiply their complementary powers together, by mixing them, before they attain superacid status. They're strongest only under contrived circumstances. Really, the strongest solo acid is still the boron-based carborane ($HCB_{11}Cl_{11}$). And this boron acid has the best punch line so far: It's simultaneously the world's strongest *and gentlest* acid. To wrap your head around that, remember that acids split into positive and negative parts. In carborane's case, you get H^+ and an elaborate cagelike structure formed by everything else ($CB_{11}Cl_{11}^-$). With most acids it's the negative portion that's corrosive and caustic and eats through skin. But the boron cage forms one of the most stable molecules ever invented. Its boron atoms share electrons so generously that it practically becomes helium, and it won't go around ripping electrons from other atoms, the usual cause of acidic carnage.

So what's carborane good for, if not dissolving glass bottles or eating through bank vaults? It can add an octane kick to gasoline, for one thing, and help make vitamins digestible. More important is its use in chemical "cradling." Many chemical reactions involving protons aren't clean, quick swaps. They require multiple steps, and protons get shuttled around in millionths of billionths of seconds—so quickly scientists have no idea what really happened. Carborane, though, because it's so stable and unreactive, will flood a solution with protons, then freeze the molecules at crucial intermediate points. Carborane holds the intermediate species up on a soft, safe pillow. In contrast, antimony superacids make terrible cradles, because they shred the molecules scientists most want to look at. Lewis would have enjoyed seeing this and other applications of his work with electrons and acids, and it might have brightened the last dark years of his life. Although he did government work during World War I and made valuable contributions to chemistry until he was in his sixties, he was passed over for the Manhattan Project during World War II. This galled him, since many chemists he had recruited to Berkeley played important roles in building the first atomic bomb and became national heroes. In contrast, he puttered around during the war, reminiscing and writing a wistful pulp novel about a soldier. He died alone in his lab in 1946.

There's general agreement that after smoking twenty-some cigars per day for forty-plus years, Lewis died of a heart attack. But it was hard not to notice that his lab smelled like bitter almonds—a sign of cyanide gas—the afternoon he died. Lewis used cyanide in his research, and it's possible he dropped a canister of it after going into cardiac arrest. Then again, Lewis had had lunch earlier in the day—a lunch he'd initially refused to attend—with a younger, more charismatic rival chemist who had won the Nobel Prize and served as a special consultant to the Manhattan Project. It's always been in the back of some people's minds that the honored colleague might have unhinged Lewis. If that's true, his facility with chemistry might have been both convenient and unfortunate.

In addition to reactive metals on its west coast and halogens and noble gases up and down its east coast, the periodic table contains a "great plains" that stretches right across its middle— columns three through twelve, the transition metals. To be honest, the transition metals have exasperating chemistry, so it's hard to say anything about them generally—except be careful. You see, heavier atoms like the transition metals have more flexibility than other atoms in how they store their electrons. Like other atoms, they have different energy levels (designated one, two, three, etc.), with lower energy levels buried beneath higher levels. And they also fight other atoms to secure full outer energy levels with eight electrons. Figuring out what counts as the outer level, however, is trickier.

As we move horizontally across the periodic table, each element has one more electron than its neighbor to the left. Sodium, element eleven, normally has eleven electrons; magnesium, element twelve, has twelve electrons; and so on. As elements swell in size, they not only sort electrons into energy levels, they also store those electrons in different-shaped bunks, called shells. But atoms, being unimaginative and conformist, fill shells and energy levels in the same order as we move across the table. Elements on the far left-hand side of the table put the first electron in an s-shell, which is spherical. It's small and holds only two electrons—which explains the two taller columns on the left side. After those first two electrons, atoms look for something roomier. Jumping across the gap, elements in the columns on the right-hand side begin to pack new electrons one by one into a p-shell, which looks like a misshapen lung. P-shells can hold six electrons, hence the six taller columns on the right. Notice that across each row near the top, the two s-shell electrons plus the six p-shell electrons add up to eight electrons total, the number most atoms want in the outer shell. And except for the self-satisfied noble gases, all these elements' outer-shell electrons are available to dump onto or react with other atoms. These elements behave in a logical manner: add a new electron, and the atom's behavior should change, since it has more electrons available to participate in reactions.

Now for the frustrating part. The transition metals appear in columns three through twelve of the fourth through seventh rows, and they start to file electrons into what are called d-shells, which hold ten electrons. (D-shells look like nothing so much as misshapen balloon animals.) Based on what every other previous element has done with its shells, you'd expect the transition metals to put each extra d-shell electron on display in an outer layer and for that extra electron to be available for reactions, too. But no, transition metals squirrel their extra electrons away and prefer to hide them beneath other layers. The decision of the transition metals to violate convention and bury their d-shell electrons seems ungainly and counterintuitive—Plato would not have liked it. It's also how nature works, and there's not much we can do about it.

There's a payoff to understanding this process. Normally as we move horizontally across the table, the addition of one electron to each transition metal would alter its behavior, as happens with elements in other parts of the table. But because the metals bury their d-shell electrons in the equivalent of false-bottomed drawers, those electrons end up shielded. Other atoms trying to react with the metals cannot get at those electrons, and the upshot is that many metals in a row leave the same number of electrons exposed. They therefore act the same way chemically. That's why, scientifically, many metals look so indistinguishable and act so indistinguishably. They're all cold, gray lumps because their outer electrons leave them no choice but to conform. (Of course, just to confuse things, sometimes buried electrons do rise up and react. That's what causes the slight differences between some metals. That's also why their chemistry is so exasperating.)

F-shell elements are similarly messy. F-shells begin to appear in the first of the two free-floating rows of metals beneath the periodic table, a group called the lanthanides. (They're also called the rare earths, and according to their atomic numbers, fifty-seven through seventy-one, they really belong in the sixth row. They were relegated to the bottom to make the table skinnier and less unwieldy.) The lanthanides bury new electrons even more deeply than the transition metals, often two energy levels down. This means they are even more alike than the transition metals and can barely be distinguished from one another. Moving along the row is like driving from Nebraska to South Dakota and not realizing you've crossed the state line.

It's impossible to find a pure sample of a lanthanide in

nature, since its brothers always contaminate it. In one famous case, a chemist in New Hampshire tried to isolate thulium, element sixty-nine. He started with huge casserole dishes of thulium-rich ore and repeatedly treated the ore with chemicals and boiled it, a process that purified the thulium by a small fraction each time. The dissolving took so long that he could do only one or two cycles per day at first. Yet he repeated this tedious process fifteen thousand times, by hand, and winnowed the hundreds of pounds of ore down to just ounces before the purity satisfied him. Even then, there was still a little cross-contamination from other lanthanides, whose electrons were buried so deep, there just wasn't enough of a chemical handle to grasp them and pull them out.

Electron behavior drives the periodic table. But to really understand the elements, you can't ignore the part that makes up more than 99 percent of their mass—the nucleus. And whereas electrons obey the laws of the greatest scientist never to win the Nobel Prize, the nucleus obeys the dictates of probably the most unlikely Nobel laureate ever, a woman whose career was even more nomadic than Lewis's.

Maria Goeppert was born in Germany in 1906. Even though her father was a sixth-generation professor, Maria had trouble convincing a Ph.D. program to admit a woman, so she bounced from school to school, taking lectures wherever she could. She finally earned her doctorate at the University of Hannover, defending her thesis in front of professors she'd never met. Not surprisingly, with no recommendations or connections, no university would hire her upon her graduation. She could enter science only obliquely, through her husband, Joseph Mayer, an American chemistry professor visiting Germany. She returned to Baltimore with him in 1930, and the newly named Goeppert-Mayer began tagging along with Mayer to work and conferences. Unfortunately, Mayer lost his job several times during the Great Depression, and the family drifted to universities in New York and then Chicago.

Most schools tolerated Goeppert-Mayer's hanging around to chat science. Some even condescended to give her work, though they refused to pay her, and the topics were stereotypically "feminine," such as figuring out what causes colors. After the Depression lifted, hundreds of her intellectual peers gathered for the Manhattan Project, perhaps the most vitalizing exchange of scientific ideas ever. Goeppert-Mayer received an invitation to participate, but peripherally, on a useless side project to separate uranium with flashing lights. No doubt she chafed in private, but she craved science enough to continue to work under such conditions. After World War II, the University of Chicago finally took her seriously enough to make her a professor of physics. Although she got her own office, the department still didn't pay her.

Nevertheless, bolstered by the appointment, she began work in 1948 on the nucleus, the core and essence of an atom. Inside the nucleus, the number of positive protons—the atomic number—determines the atom's identity. In other words, an atom cannot gain or lose protons without becoming a different element. Atoms do not normally lose neutrons either, but an element's atoms can have different numbers of neutrons—variations called isotopes. For instance, the isotopes lead-204 and lead-206 have identical atomic numbers (82) but different numbers of neutrons (122 and 124). The atomic number plus the number of neutrons is called the atomic weight. It took scientists many years to figure out the relationship between atomic number and atomic weight, but once they did, periodic table science got a lot clearer.

Goeppert-Mayer knew all this, of course, but her work touched on a mystery that was more difficult to grasp, a deceptively simple problem. The simplest element in the universe, hydrogen, is also the most abundant. The second-simplest element, helium, is the second most abundant. In an aesthetically tidy universe, the third element, lithium, would be the third most abundant, and so on. Our universe isn't tidy. The third most common element is oxygen, element eight. But why? Scientists might answer that oxygen has a very stable nucleus, so it doesn't disintegrate, or "decay." But that only pushed the question back—why do certain elements like oxygen have such stable nuclei?

Unlike most of her contemporaries, Goeppert-Mayer saw a parallel here to the incredible stability of noble gases. She suggested that protons and neutrons in the nucleus sit in shells just like electrons and that filling nuclear shells leads to stability. To an outsider, this seems reasonable, a nice analogy. But Nobel Prizes aren't won on conjectures, especially those by unpaid female professors. What's more, this idea ruffled nuclear scientists, since chemical and nuclear processes are independent. There's no reason why dependable, stay-at-home neutrons and protons should behave like tiny, capricious electrons, which abandon their homes for attractive neighbors. And mostly they don't.

Except Goeppert-Mayer pursued her hunch, and by piecing together a number of unlinked experiments, she proved that nuclei do have shells and do form what she called magic nuclei. For complex mathematical reasons, magic nuclei don't reappear periodically like elemental properties. The magic happens at atomic numbers two, eight, twenty, twenty-eight, fifty, eighty-two, and so on. Goeppert-Mayer's work proved how, at those numbers, protons and neutrons marshal themselves into highly stable, highly symmetrical spheres. Notice too that oxygen's eight protons and eight neutrons make it doubly magic and therefore eternally stable—which explains its seeming over-abundance. This model also explains at a stroke why elements such as calcium (twenty) are disproportionately plentiful and, not incidentally, why our bodies employ these readily available minerals.

Goeppert-Mayer's theory echoes Plato's notion that beautiful shapes are more perfect, and her model of magic, orb-shaped nuclei became the ideal form against which all nuclei are judged. Conversely, elements stranded far between two magic numbers are less abundant because they form ugly, oblong nuclei. Scientists have even discovered neutron-starved forms of holmium (element sixty-seven) that give birth to a deformed, wobbly "football nucleus." As you might guess from Goeppert-Mayer's model (or from ever having watched somebody fumble during a football game), the holmium footballs aren't very steady. And unlike atoms with misbalanced electron shells, atoms with distorted nuclei can't poach neutrons and protons from other atoms to balance themselves. So atoms with misshapen nuclei, like that form of holmium, hardly ever form and immediately disintegrate if they do.

The nuclear shell model is brilliant physics. That's why it no doubt dismayed Goeppert-Mayer, given her precarious status among scientists, to discover that it had been duplicated by male physicists in her homeland. She risked losing credit for everything. However, both sides had produced the idea independently, and when the Germans graciously acknowledged her work and asked her to collaborate, Goeppert-Mayer's career took off. She won her own accolades, and she and her husband moved a final time in 1959, to San Diego, where she began a real, paying job at the new University of California campus there. Still, she never quite shook the stigma of being a dilettante. When the Swedish Academy announced in 1963 that she had won her profession's highest honor, the San Diego newspaper greeted her big day with the headline "S.D. Mother Wins Nobel Prize."

But maybe it's all a matter of perspective. Newspapers could have run a similarly demeaning headline about Gilbert Lewis, and he probably would have been thrilled.

Reading the periodic table across each row reveals a lot about the elements, but that's only part of the story, and not even the best part. Elements in the same column, latitudinal neighbors, are actually far more intimately related than horizontal neighbors. People are used to reading from left to right (or right to left) in virtually every human language, but reading the periodic table up and down, column by column, as in some forms of Japanese, is actually more significant. Doing so reveals a rich subtext of relationships among elements, including unexpected rivalries and antagonisms. The periodic table has its own grammar, and reading between its lines reveals whole new stories.

2

Near Twins and Black Sheep: The Genealogy of Elements

$\begin{array}{|c|}\hline C \\ \hline \end{array}$ 6 12.011 $\begin{array}{|c|}\hline Si \\ \hline \end{array}$ 14 28.086 $\begin{array}{|c|}\hline Ge \\ \hline \end{array}$ 32 72.641

Shakespeare had a go at it with "honorificabilitudinitati-bus"—which, depending on whom you ask, either means "the state of being loaded with honors" or is an anagram proclaiming that Francis Bacon, not the Bard, really wrote Shakespeare's plays.* But that word, a mere twenty-seven letters, doesn't stretch nearly long enough to count as the longest word in the English language.

Of course, determining *the* longest word is like trying to wade into a riptide. You're likely to lose control quickly, since language is fluid and constantly changing direction. What even qualifies as English differs in different contexts. Shakespeare's word, spoken by a clown in *Love's Labor's Lost*, obviously comes from Latin. But perhaps foreign words, even in English sentences, shouldn't count. Plus, if you count words that do little but stack suffixes and prefixes together ("antidisestablishmentarianism," twenty-eight letters) or nonsense words ("supercalifragilisticexpialidocious," thirty-four letters), writers can string readers along pretty much until their hands cramp up.

But if we adopt a sensible definition—the longest word to

appear in an English-language document whose purpose was *not* to set the record for the longest word ever—then the word we're after appeared in 1964 in *Chemical Abstracts*, a dictionary-like reference source for chemists. The word describes an important protein on what historians generally count as the first virus ever discovered, in 1892—the tobacco mosaic virus. Take a breath.

acetylseryltyrosylserylisoleucylthreonylserylprolylseryl-glutaminylphenylalanylvalylphenylalanylleucylserylseryl-valyltryptophylalanylaspartylprolylisoleucylglutamyl-leucylleucylasparaginylvalylcysteinylthreonylserylseryl-leucylglycylasparaginylglutaminylphenylalanylglutami-nylthreonylglutaminylglutaminylalanylarginylthreo-nylthreonylglutaminylvalylglutaminylglutaminylpheny-lalanylserylglutaminylvalyltryptophyllysylprolylphenyla-lanylprolylglutaminylserylthreonylvalylarginylphenylala-nylprolylglycylaspartylvalyltyrosyllysylvalyltyrosylargin-yltyrosylasparaginylalanylvalylleucylaspartylprolylleucyli-soleucylthreonylalanylleucylleucylglycylthreonylphenyla-lanylaspartylthreonylarginylasparaginylarginylisoleucyli-soleucylglutamylvalylglutamylasparaginylglutaminylglu-taminylserylprolylthreonylthreonylalanylglutamylthreo-nylleucylaspartylalanylthreonylarginylarginylvalylaspar-tylaspartylalanylthreonylvalylalanylisoleucylarginylsery-lalanylasparaginylisoleucylasparaginylleucylvalylasparagi-nylglutamylleucylvalylarginylglycylthreonylglycylleucyl-tyrosylasparaginylglutaminylasparaginylthreonylphenyla-lanylglutamylserylmethionylserylglycylleucylvalyltrypto-phylthreonylserylalanylprolylalanylserine

That anaconda runs 1,185 letters.*

Now, since none of you probably did more than run your eyes across "acetyl...serine," go back and take a second look. You'll notice something funny about the distribution of letters. The most common letter in English, *e*, appears 65 times; the uncommon letter *y* occurs 183 times. One letter, *l*, accounts for 22 percent of the word (255 instances). And the *y* and *l* don't appear randomly but often next to each other—166 pairs, every seventh letter or so. That's no coincidence. This long word describes a protein, and proteins are built up from the sixth (and most versatile) element on the periodic table, carbon.

Specifically, carbon forms the backbone of amino acids, which string together like beads to form proteins. (The tobacco mosaic virus protein consists of 159 amino acids.) Biochemists, because they often have so many amino acids to count, catalog them with a simple linguistic rule. They truncate the *ine* in amino acids such as "serine" or "isoleucine" and alter it to *yl*, making it fit a regular meter: "seryl" or "isoleucyl." Taken in order, these linked *yl* words describe a protein's structure precisely. Just as laypeople can see the compound word "matchbox" and grasp its meaning, biochemists in the 1950s and early 1960s gave molecules official names like "acetyl...serine" so they could reconstruct the whole molecule from the name alone. The system was exact, if exhausting. Historically, the tendency to amalgamate words reflects the strong influence that Germany and the compound-crazy German language had on chemistry.

But why do amino acids bunch together in the first place? Because of carbon's place on the periodic table and its need to fill its outer energy level with eight electrons—a rule of thumb called the octet rule. On the continuum of how aggressively atoms and molecules go after one another, amino acids shade toward the more civilized end. Each amino acid contains oxygen atoms on one end, a nitrogen on the other, and a trunk of two carbon atoms in the middle. (They also contain hydrogen and a branch off the main trunk that can be twenty different molecules, but those don't concern us.) Carbon, nitrogen, and oxygen all want to get eight electrons in the outer level, but it's easier for one of these elements than for the other. Oxygen, as element eight, has eight total electrons. Two belong to the lowest energy tier, which fills first. That leaves six left over in the outer level, so oxygen is always scouting for two additional electrons. Two electrons aren't so hard to find, and aggressive oxygen can dictate its own terms and bully other atoms. But the same arithmetic shows that poor carbon, element six, has four electrons left over after filling its first shell and therefore needs four more to make eight. That's harder to do, and the upshot is that carbon has really low standards for forming bonds. It latches onto virtually anything.

That promiscuity is carbon's virtue. Unlike oxygen, carbon must form bonds with other atoms in whatever direction it can. In fact, carbon shares its electrons with up to four other atoms at once. This allows carbon to build complex chains, or even three-dimensional webs of molecules. And because it shares and cannot steal electrons, the bonds it forms are steady and stable. Nitrogen also must form multiple bonds to keep itself happy, though not to the same degree as carbon. Proteins like that anaconda described earlier simply take advantage of these elemental facts. One carbon atom in the trunk of an amino acid shares an electron with a nitrogen at the butt of another, and proteins arise when these connectible carbons and nitrogens are strung along pretty much ad infinitum, like letters in a very, very long word.

In fact, scientists nowadays can decode vastly longer molecules than "acetyl...serine." The current record is a gargantuan protein whose name, if spelled out, runs 189,819 letters. But during the 1960s, when a number of quick amino acid sequencing tools became available, scientists realized that they would soon end up with chemical names as long as this book (the spell-checking of which would have been a real bitch). So they dropped the unwieldy Germanic system and reverted to shorter, less bombastic titles, even for official purposes. The 189,819-letter molecule, for instance, is now mercifully known as titin.* Overall, it seems doubtful that anyone will ever top the mosaic virus protein's full name in print, or even try.

That doesn't mean aspiring lexicographers shouldn't still brush up on biochemistry. Medicine has always been a fertile source of ridiculously long words, and the longest nontechnical word in the *Oxford English Dictionary* just happens to be based on the nearest chemical cousin of carbon, an element often cited as an alternative to carbon-based life in other galaxies—element fourteen, silicon.

In genealogy, parents at the top of a family tree produce children who resemble them, and in just the same way, carbon has more in common with the element below it, silicon, than with its two horizontal neighbors, boron and nitrogen. We already know the reason. Carbon is element six and silicon element fourteen, and that gap of eight (another octet) is not coincidental. For silicon, two electrons fill the first energy level, and eight fill the second. That leaves four more electrons—and leaves silicon in the same predicament as carbon. But being in that situation gives silicon some of carbon's flexibility, too. And because carbon's flexibility is directly linked to its capacity to form life, silicon's ability to mimic carbon has made it the dream of generations of science fiction fans interested in alternative—that is, alien—modes of life, life that follows different rules than earth-bound life. At the same time, genealogy isn't destiny, since children are never exactly like their parents. So while carbon and silicon are indeed closely related, they're distinct elements that form distinct compounds. And unfortunately for science fiction fans, silicon just can't do the wondrous tricks carbon can.

Oddly enough, we can learn about silicon's limitations by parsing another record-setting word, a word that stretches a ridiculous length for the same reason the 1,185-letter carbon-based protein above did. Honestly, that protein has sort of a formulaic name—interesting mostly for its novelty, the same way that calculating pi to trillions of digits is. In contrast, the longest nontechnical word in the *Oxford English Dictionary* is the forty-five-letter "pneumonoultramicroscopicsilicovolcanoconiosis," a disease that has "silico" at its core. Logologists (word nuts) slangily refer to pneumonoultramicroscopicsilicovolcanoconiosis as "p45," but there's some medical question about whether p45 is a real disease, since it's just a variant of an incurable lung condition called pneumonoconiosis. P16 resembles pneumonia and is one of the diseases that inhaling asbestos causes. Inhaling silicon dioxide, the major component of sand and glass, can cause pneumonoconiosis, too. Construction workers who sandblast all day and insulation plant assembly-line workers who inhale glass dust often come down with silicon-based p16. But because silicon dioxide (SiO_2) is the most common mineral in the earth's crust, one other group is susceptible: people who live in the vicinity of active volcanoes. The most powerful volcanoes pulverize silica into fine bits and spew megatons of it into the air. Those bits are prone to wriggling into lung sacs. Because our lungs regularly deal with carbon dioxide, they see nothing wrong with absorbing its cousin, SiO_2, which can be fatal. Many dinosaurs might have died this way when a metropolis-sized asteroid or comet struck the earth 65 million years ago.

With all that in mind, parsing the prefixes and suffixes of p45 should now be a lot easier. The lung disease caused by inhaling fine volcanic silica as people huff and puff to flee the scene is naturally called pneumono-ultra-microscopic-silico-volcano-coniosis. Before you start dropping it in conversation, though, know that many word purists detest it. Someone coined p45 to win a puzzle contest in 1935, and some people still sneer that it's a "trophy word." Even the august editors of the *Oxford English Dictionary* malign p45 by defining it as "a fractious word," one that is only "alleged to mean" what it does. This loathing arises because p45 just expanded on a "real" word. P45 was tinkered with, like artificial life, instead of rising organically from everyday language.

By digging further into silicon, we can explore whether claims of silicon-based life are tenable. Though as overdone a trope in science fiction as ray guns, silicon life is an important idea because it expands on our carbon-centric notion of life's potential. Silicon enthusiasts can even point to a few animals on earth that employ silicon in their bodies, such as sea urchins with their silicon spines and radiolarian protozoa (one-celled creatures) that forge silicon into exoskeletal armor. Advances in computing and artificial intelligence also suggest that silicon could form "brains" as complicated as any carbon-based one. In theory, there's no reason you couldn't replace every neuron in your brain with a silicon transistor.

But p45 provides lessons in practical chemistry that dash hopes for silicon life. Obviously silicon life forms would need to shuttle silicon into and out of their bodies to repair tissues or whatever, just as earth-based creatures shuttle carbon around. On earth, creatures at the base of the food chain (in many ways, the most important forms of life) can do that via gaseous carbon dioxide. Silicon almost always bonds to oxygen in nature, too, usually as SiO_2. But unlike carbon dioxide, silicon dioxide (even as fine volcanic dust) is a solid, not a gas, at any temperature remotely friendly to life. (It doesn't become a gas until 4,000°F!) On the level of cellular respiration, breathing solids just doesn't work, because solids stick together. They don't flow, and it's hard to get at individual molecules, which cells need to do. Even rudimentary silicon life, the equivalent of pond scum, would have trouble breathing, and larger life forms with multiple layers of cells would be even worse off. Without ways to exchange gases with the environment, plant-like silicon life would starve and animal-like silicon life would suffocate on waste, just like our carbon-based lungs are smothered by p45.

Couldn't those silicon microbes expel or suck up silica in other ways, though? Possibly, but silica doesn't dissolve in water, the most abundant liquid in the universe by far. So those creatures would have to forsake the evolutionary advantages of blood or any liquid to circulate nutrients and waste. Silicon-based creatures would have to rely on solids, which don't mix easily, so it's impossible to imagine silicon life forms *doing* much of anything.

Furthermore, because silicon packs on more electrons than carbon, it's bulkier, like carbon with fifty extra pounds. Sometimes that's not a big deal. Silicon might substitute adequately for carbon in the Martian equivalent of fats or proteins. But carbon also contorts itself into ringed molecules we call sugars. Rings are states of high tension—which means they store lots of energy—and silicon just isn't supple enough to bend into the right position to form rings. In a related problem, silicon atoms cannot squeeze their electrons into tight spaces for double bonds, which appear in virtually every complicated biochemical. (When two atoms share two electrons, that's a single bond. Sharing four electrons is a double bond.) Silicon-based life would therefore have hundreds of fewer options for storing chemical energy and making chemical hormones. Altogether, only a radical biochemistry could support silicon life that actually grows, reacts, reproduces, and attacks. (Sea urchins and radiolaria use silica only for structural support, not for breathing or storing energy.) And the fact that carbon-based life evolved on earth despite carbon being vastly less common than silicon is almost a proof in itself.* I'm not foolish enough to predict that silicon biology is impossible, but unless those creatures defecate sand and live on planets with volcanoes constantly expelling ultramicroscopic silica, this element probably isn't up to the task of making life live.

Luckily for it, silicon has ensured itself immortality in another way. Like a virus, a quasi-living creature, it wriggled into an evolutionary niche and has survived by preying parasitically on the element below it.

There are further genealogical lessons in carbon and silicon's column of the periodic table. Under silicon, we find germanium. One element down from germanium, we unexpectedly find tin. One space below that is lead. Moving straight down the periodic table, then, we pass from carbon, the element responsible for life; to silicon and germanium, elements responsible for modern electronics; to tin, a dull gray metal used to can corn; to lead, an element more or less hostile to life. Each step is small, but it's a good reminder that while an element may resemble the one below it, small mutations accumulate.

Another lesson is that every family has a black sheep, someone the rest of the line more or less has given up on. In column fourteen's case, it's germanium, a sorry, no-luck element. We use silicon in computers, in microchips, in cars and calculators. Silicon semiconductors sent men to the moon and drive the Internet. But if things had gone differently sixty years ago, we might all be talking about Germanium Valley in northern California today.

The modern semiconductor industry began in 1945 at Bell Labs in New Jersey, just miles from where Thomas Alva Edison set up his invention factory seventy years before. William Shockley, an electrical engineer and physicist, was trying to build a small silicon amplifier to replace vacuum tubes in mainframe computers. Engineers loathed vacuum tubes because the long, lightbulb-like glass shells were cumbersome, fragile, and prone to overheating. Despise them as they may, they needed these tubes, because nothing else could pull their double duty: the tubes both amplified electronic signals, so faint signals didn't die, and acted as one-way gates for electricity, so electrons couldn't flow backward in circuits. (If your sewer pipes flowed both ways, you can imagine the potential problems.) Shockley set out to do to vacuum tubes what Edison had done to candles, and he knew that semiconducting elements were the answer: only they could achieve the balance engineers wanted by letting enough electrons through to run a circuit (the "conductor" part), but not so many that the electrons were impossible to control (the "semi" part). Shockley, though, was more visionary than engineer, and his silicon amplifier never amplified anything. Frustrated after two unfruitful years, he dumped the task onto two underlings, John Bardeen and Walter Brattain.

Bardeen and Brattain, according to one biographer, "loved one another as much as two men can.... It was like Bardeen was the brains of this joint organism and Brattain was the hands."* This symbiosis was convenient, since Bardeen, for whom the descriptor "egghead" might have been coined, wasn't so adept with his own hands. The joint organism soon determined that silicon was too brittle and difficult to purify to work as an amp. Plus, they knew that germanium, whose outer electrons sit in a higher energy level than silicon's and therefore are more loosely bound, conducted electricity more smoothly. Using germanium, Bardeen and Brattain built the world's first solid-state (as opposed to vacuum) amplifier in December 1947. They called it the transistor.

This should have thrilled Shockley—except he was in Paris that Christmas, making it hard for him to claim he'd contributed to the invention (not to mention that he had used the wrong element). So Shockley set out to steal credit for Bardeen and Brattain's work. Shockley wasn't a wicked man, but he was ruthless when convinced he was right, and he was convinced he deserved most of the credit for the transistor. (This ruthless conviction resurfaced later in Shockley's declining years, after he abandoned solid-state physics for the "science" of eugenics—the breeding of better human beings. He believed in a Brahmin caste of intelligentsia, and he began donating to a "genius sperm bank"* and advocating that poor people and minoritics be paid to get sterilized and stop diluting humankind's collective IQ.)

Hurrying back from Paris, Shockley wedged himself back into the transistor picture, often literally. In Bell Labs publicity photos showing the three men supposedly at work, he's always standing between Bardeen and Brattain, dissecting the joint organism and putting *his* hands on the equipment, forcing the other two to peer over his shoulders like mere assistants. Those images became the new reality and the general scientific community gave credit to all three men. Shockley also, like a petty prince in a fiefdom, banished his main intellectual rival, Bardeen, to another, unrelated lab so that he, Shockley, could develop a second and more commercially friendly generation of germanium transistors. Unsurprisingly, Bardeen soon quit Bell Labs to take an academic post in Illinois. He was so disgusted, in fact, that he gave up semiconductor research.

Things turned sour for germanium, too. By 1954, the transistor industry had mushroomed. The processing power of computers had expanded by orders of magnitude, and whole new product lines, such as pocket radios, had sprung up. But throughout the boom, engineers kept ogling silicon. Partly they did so because germanium was temperamental. As a corollary of conducting electricity so well, it generated unwanted heat, too, causing germanium transistors to stall at high temperatures. More important, silicon, the main component of sand, was perhaps even cheaper than proverbial dirt. Scientists were still faithful to germanium, but they were spending an awful lot of time fantasizing about silicon.

Suddenly, at a semiconductor trade meeting that year, a cheeky engineer from Texas got up after a gloomy speech about the unfeasibility of silicon transistors and announced that he actually had one in his pocket. Would the crowd like a demonstration? This P. T. Barnum—whose real name was Gordon Teal—then hooked up a germanium-run record player to external speakers and, rather medievally, lowered the player's innards into a vat of boiling oil. As expected, it choked and died. After fishing the innards out, Teal popped out the germanium transistor and rewired the record player with his silicon one. Once again, he plopped it into the oil. The band played on. By the time the stampede of salesmen reached the pay phones at the back of the convention hall, germanium had been dumped.

Luckily for Bardeen, his part of the story ended happily, if clumsily. His work with germanium semiconductors proved so important that he, Brattain, and, *sigh*, Shockley all won the

Nobel Prize in physics in 1956. Bardeen heard the news on his radio (by then probably silicon-run) while frying breakfast one morning. Flustered, he knocked the family's scrambled eggs onto the floor. It was not his last Nobel-related gaffe. Days before the prize ceremony in Sweden, he washed his formal white bow tie and vest with some colored laundry and stained them green, just as one of his undergrad students might have. And on the day of the ceremony, he and Brattain got so wound up about meeting Sweden's King Gustav I that they chugged quinine to calm their stomachs. It probably didn't help when Gustav chastened Bardeen for making his sons stay in class back at Harvard (Bardeen was afraid they might miss a test) instead of coming to Sweden with him. At this rebuke, Bardeen tepidly joked that, ha, ha, he'd bring them along the next time he won the Nobel Prize.

Gaffes aside, the ceremony marked a high point for semiconductors, but a brief one. The Swedish Academy of Sciences, which hands out the Nobel Prizes in Chemistry and Physics, tended at the time to honor pure research over engineering, and the win for the transistor was an uncommon acknowledgment of applied science. Nevertheless, by 1958 the transistor industry faced another crisis. And with Bardeen out of the field, the door stood open for another hero.

Although he probably had to stoop (he stood six feet six), Jack Kilby soon walked through it. A slow-talking Kansan with a leathery face, Kilby had spent a decade in the high-tech boondocks (Milwaukee) before landing a job at Texas Instruments (TI) in 1958. Though trained in electrical engineering, Kilby was hired to solve a computer hardware problem known as the tyranny of numbers. Basically, though cheap silicon transistors worked okay, fancy computer circuits required scores of them. That meant companies like TI had to employ whole hangars of low-paid, mostly female technicians who did nothing all day but crouch over microscopes, swearing and sweating in hazmat suits, as they soldered silicon bits together. In addition to being expensive, this process was inefficient. In every circuit, one of those frail wires inevitably broke or worked loose, and the whole circuit died. Yet engineers couldn't get around the need for so many transistors: the tyranny of numbers.

Kilby arrived at TI during a sweltering June. As a new employee he had no vacation time, so when the cast of thousands cleared out for mandatory vacations in July, he was left alone at his bench. The relief of silence no doubt convinced him that employing thousands of people to wire transistors together was asinine, and the absence of supervisors gave him free time to pursue a new idea he called an integrated circuit. Silicon transistors weren't the only parts of a circuit that had to be hand-wired. Carbon resistors and porcelain capacitors also had to be spaghettied together with copper wire. Kilby scrapped that separate-element setup and instead carved everything—all the resistors, transistors, and capacitors—from one firm block of semiconductor. It was a smashing idea—the difference, structurally and artistically, between sculpting a statue from one block of marble and carving each limb separately, then trying to fit the statue together with wire. Not trusting the purity of silicon to make the resistors and capacitors, he turned to germanium for his prototype.

Ultimately, this integrated circuit freed engineers from the tyranny of hand-wiring. Because the pieces were all made of the same block, no one had to solder them together. In fact, soon no one even could have soldered them together, because the integrated circuit also allowed engineers to automate the carving process and make microscopic sets of transistors—the first real computer chips. Kilby never received full credit for his innovation (one of Shockley's protégés filed a rival, and slightly more detailed, patent claim a few months later and wrestled the rights away from Kilby's company), but geeks today still pay Kilby the ultimate engineering tribute. In an industry that measures product cycles in months, chips are still made using his basic design fifty years later. And in 2000, he won a belated Nobel Prize for his integrated circuit.*

Sadly, though, nothing could resurrect germanium's reputation. Kilby's original germanium circuit is ensconced in the Smithsonian Institution, but in the bare-knuckle marketplace, germanium got pummeled. Silicon was too cheap and too available. Sir Isaac Newton famously said that he had achieved everything by standing on the shoulders of giants—the scientific men whose findings he built upon. The same might be said about silicon. After germanium did all the work, silicon became an icon, and germanium was banished to periodic table obscurity.

In truth, that's a common fate regarding the periodic table. Most elements are undeservedly anonymous. Even the names of the scientists who discovered many of them and who arranged them into the first periodic tables have long since been forgotten. Yet like silicon, a few names have achieved universal fame, and not always for the best reasons. All of the scientists working on early periodic tables recognized likenesses among certain elements. Chemical "triads," like the modern-day example of carbon, silicon, and germanium, were the first clue that the periodic system existed. But some scientists proved more facile than others at recognizing subtleties—the traits that run through the families of the periodic table like dimples or crooked noses in humans. Knowing how to trace and predict such similarities soon allowed one scientist, Dmitri Mendeleev, to vault into history as the father of the periodic table.

3

The Galápagos of the Periodic Table

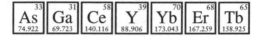

Y ou might say the history of the periodic table is the history of the many characters who shaped it. The first had one of those names from history books, like Dr. Guillotin, or Charles Ponzi, or Jules Léotard, or Étienne de Silhouette, that makes you smile to think someone actually answered to it. This pioneer of the periodic table deserves special praise, since his eponymous burner has enabled more sophomoric stunts than any lab equipment in history. Disappointingly, German chemist Robert Bunsen didn't actually invent "his" burner, just improved the design and popularized it in the mid-1800s. Even without the Bunsen burner, he managed to pack plenty of danger and destruction into his life.

Bunsen's first love was arsenic. Although element thirty-three has had quite a reputation since ancient times (Roman assassins used to smear it on figs), few law-abiding chemists knew much about arsenic before Bunsen started sloshing it around in test tubes. He worked primarily with arsenic-based cacodyls, chemicals whose name is based on the Greek word for "stinky." Cacodyls smelled so foul, Bunsen said, they made him hallucinate, "produc[ing] instantaneous tingling of the hands and feet, even giddiness and insensibility." His tongue became "covered with a black coating." Perhaps from self-interest, he soon developed what's still the best antidote to arsenic poisoning, iron oxide hydrate, a chemical related to rust that clamps onto arsenic in the blood and drags it out. Still, he couldn't shield himself from every danger. The careless explosion of a glass beaker of arsenic nearly blew out his right eye and left him half-blind for the last sixty years of his life.

After the accident, Bunsen put arsenic aside and indulged his passion for natural explosions. Bunsen loved anything that spewed from the ground, and for several years he investigated geysers and volcanoes by hand-collecting their vapors and boiling liquids. He also jury-rigged a faux Old Faithful in his laboratory and discovered how geysers build up pressure and blow. Bunsen settled back into chemistry at the University of Heidelberg in the 1850s and soon ensured himself scientific immortality by inventing the spectroscope, which uses light to study elements. Each element on the periodic table produces sharp, narrow bands of colored light when heated. Hydrogen, for example, always emits one red, one yellowish green, one baby blue, and one indigo band. If you heat some mystery substance and it emits those specific lines, you can bet it contains hydrogen. This was a powerful breakthrough, the first way to peer inside exotic compounds without boiling them down or disintegrating them with acid.

To build the first spectroscope, Bunsen and a student mounted a prism inside a discarded cigar box, to keep out stray light, and attached two broken-off eyepieces from telescopes to peer inside, like a diorama. The only thing limiting spectroscopy at that point was getting flames hot enough to excite elements. So Bunsen duly invented the device that made him a hero to everyone who ever melted a ruler or started a pencil on fire. He took a local technician's primitive gas burner and added a valve to adjust the oxygen flow. (If you remember fussing around with the knob on the bottom of your Bunsen burner, that's it.) As a result, the burner's flame improved from an inefficient, crackling orange to the tidy, hissing blue you see on good stoves today.

Bunsen's work helped the periodic table develop rapidly. Although he opposed the idea of classifying elements by their spectra, other scientists had fewer qualms, and the spectroscope immediately began identifying new elements. Just as important, it helped sort through spurious claims by finding old elements in disguise in unknown substances. Reliable identification got chemists a long way toward the ultimate goal of understanding matter on a deeper level. Still, beyond finding new elements, scientists needed to organize them into a family tree of some sort. And here we come to Bunsen's other great contribution to the table—his help in building an intellectual dynasty in science at Heidelberg, where he instructed a number of people responsible for early work in periodic law. This includes our second character, Dmitri Mendeleev, the man generally acclaimed for creating the first periodic table.

Truth be told, like Bunsen and the burner, Mendeleev didn't conjure up the first periodic table on his own. Six people invented it independently, and all of them built on the "chemical affinities" noted by an earlier generation of chemists. Mendeleev started with a rough idea of how to group elements into small, synonymous sets, then transformed these gestures at a periodic system into scientific law, much like Homer transformed disconnected Greek myths into *The Odyssey*. Science needs heroes as much as any other field, and Mendeleev became the protagonist of the story of the periodic table for a number of reasons.

For one, he had a hell of a biography. Born in Siberia, the youngest of fourteen children, Mendeleev lost his father in 1847, when the boy was thirteen. Boldly for the time, his mother took over a local glass factory to support the family and managed the male craftsmen working there. Then the factory burned down. Pinning her hopes on her sharp-minded son, she bundled him up on horseback and rode twelve hundred miles across the steppes and steep, snowy Ural Mountains to an elite university in Moscow—which rejected Dmitri because he wasn't local stock. Undaunted, Mama Mendeleev bundled him back up and rode four hundred miles farther, to his dead father's alma mater in St. Petersburg. Just after seeing him enrolled, she died.

Mendeleev proved to be a brilliant student. After graduation, he studied in Paris and Heidelberg, where the eminent Bunsen supervised him for a spell (the two clashed personally, partly because Mendeleev was moody and partly because of Bunsen's notoriously loud and foul-fumed lab). Mendeleev returned to St. Petersburg as a professor in the 1860s and there began to think about the nature of elements, work that culminated in his famous periodic table of 1869.

Many others were working on the problem of how to organize elements, and some even solved it, however haltingly, with the same approach as Mendeleev. In England, a thirty-something chemist named John Newlands presented his make-shift table to a chemistry society in 1865. But a rhetorical blunder doomed Newlands. At the time, no one knew about the noble gases (helium through radon), so the top rows of his periodic table contained only seven units. Newlands whimsically compared the seven columns to the do-re-mi-fa-sol-la-ti-do of the musical scale. Unfortunately, the Chemical Society of London was not the most whimsical audience, and they ridiculed Newlands's nickelodeon chemistry.

The more serious rival to Mendeleev was Julius Lothar Meyer, a German chemist with an unruly white beard and neatly oiled black hair. Meyer had also worked under Bunsen at Heidelberg and had serious professional credentials. Among other things, he'd figured out that red blood cells transport oxygen by binding it to hemoglobin. Meyer published his table at practically the same time as Mendeleev, and the two even split a prestigious pre–Nobel Prize called the Davy Medal in 1882 for codiscovering the "periodic law." (It was an English prize, but Newlands was shut out until 1887, when he earned his own Davy Medal.) While Meyer continued to do great work that added to his reputation—he helped popularize a number of radical theories that turned out correct—Mendeleev turned cranky, a queer fish who, incredibly, refused to believe in the reality of atoms.* (He would later also reject other things he couldn't see, such as electrons and radioactivity.) If you had sized up the two men around 1880 and judged which was the greater theoretical chemist, you might have picked Meyer. So what separated Mendeleev from Meyer and the four other chemists who published tables before them, at least in history's judgment?*

First, more than any other chemist, Mendeleev understood that certain traits about elements persist, even if others don't. He realized a compound like mercuric oxide (an orange solid) doesn't somehow "contain" a gas, oxygen, and a liquid metal, mercury, as others believed. Rather, mercuric oxide contains two elements that happen to form a gas and a metal when separate. What stays constant is each element's atomic weight, which Mendeleev considered its defining trait, very close to the modern view.

Second, unlike others who had dabbled in arranging elements into columns and rows, Mendeleev had worked in chemistry labs his whole life and had acquired a deep, deep knowledge of how elements felt and smelled and reacted, especially metals, the most ambiguous and knotty elements to place on the table. This allowed him to incorporate all sixty-two known elements into his columns and rows. Mendeleev also revised his table obsessively, at one point writing elements on index cards and playing a sort of chemical solitaire in his office. Most important of all, while both Mendeleev and Meyer left gaps on their table where no known elements fit, Mendeleev, unlike the squeamish Meyer, had balls enough to predict that new elements would be dug up. *Look harder, you chemists and geologists*, he seemed to taunt, *and you'll find them*. By tracing the traits of known elements down each column, Mendeleev even predicted the densities and atomic weights of hidden elements, and when some predictions proved correct, people were mesmerized. Furthermore, when scientists discovered noble gases in the 1890s, Mendeleev's table passed a crucial test, since it easily incorporated the gases by adding one new column. (Mendeleev denied that noble gases existed at first, but by then the periodic table was no longer just his.)

Then there was Mendeleev's outsized character. Like his Russian contemporary Dostoevsky—who wrote his entire novel *The Gambler* in three weeks to pay off desperate gambling debts—Mendeleev threw together his first table to meet a textbook publisher's deadline. He'd already written volume one of the textbook, a five-hundred-page tome, but had got through just eight elements. That meant he had to fit all the rest into volume two. After six weeks of procrastinating, he decided in one inspired moment that the most concise way to present the information was in a table. Excited, he blew off his side job as a chemistry consultant for local cheese factories to compile the table. When the book appeared in print, Mendeleev not only predicted that new elements would fit into empty boxes beneath the likes of silicon and boron, but he also provisionally named them. It couldn't have hurt his reputation (people seek gurus during uncertain times) that he used an exotic, mystical language to create those names, using the Sanskrit word for *beyond:* eka-silicon, eka-boron, and so on.

A few years later, Mendeleev, now famous, divorced his wife and wanted to remarry. Although the conservative local church said he had to wait seven years, he bribed a priest and got on with the nuptials. This technically made him a bigamist, but no one dared arrest him. When a local bureaucrat complained to the tsar about the double standard applied to the case—the priest was defrocked—the tsar primly replied, "I admit, Mendeleev has two wives, but I have only one Mendeleev." Still, the tsar's patience wasn't infinite. In 1890, Mendeleev, a self-professed anarchist, was booted out of his academic post for sympathizing with violent leftist student groups.

It's easy to see why historians and scientists grew attached to Mendeleev's life's tale. Of course, no one would remember his biography today had he not constructed his periodic table. Overall, Mendeleev's work is comparable to that of Darwin in evolution and Einstein in relativity. None of those men did *all* the work, but they did the most work, and they did it more elegantly than others. They saw how far the consequences extended, and they backed up their findings with reams of evidence. And like Darwin, Mendeleev made lasting enemies for his work. Naming elements he'd never seen was presumptuous, and doing so infuriated the intellectual successor of Robert Bunsen—the man who discovered "eka-aluminium" and justifiably felt that he, not the rabid Russian, deserved credit and naming rights.

*　　*　　*

The discovery of eka-aluminium, now known as gallium, raises the question of what really drives science forward — theories, which frame how people view the world, or experiments, the simplest of which can destroy elegant theories. After a dustup with the theorist Mendeleev, the experimentalist who discovered gallium had a definite answer. Paul Emile François Lecoq de Boisbaudran was born into a winemaking family in the Cognac region of France in 1838. Handsome, with sinuous hair and a curled mustache, prone to wearing stylish cravats, he moved to Paris as an adult, mastered Bunsen's spectroscope, and became the best spectroscopic surgeon in the world.

Lecoq de Boisbaudran grew so adroit that in 1875, after spotting never-before-seen color bands in a mineral, he concluded, instantly and correctly, he'd discovered a new element. He named it gallium, after Gallia, the Latin name for France. (Conspiracy mongers accused him of slyly naming the element after himself, since Lecoq, or "the rooster," is *gallus* in Latin.) Lecoq de Boisbaudran decided he wanted to hold and feel his new prize, so he set about purifying a sample of it. It took a few years, but by 1878 the Frenchman finally had a nice, pure hunk of gallium. Though solid at moderate room temperature, gallium melts at 84°F, meaning that if you hold it in the palm of your hand (because body temperature is about 98°F), it will melt into a grainy, thick puddle of pseudoquicksilver. It's one of the few liquid metals you can touch without boiling your finger to the bone. As a result, gallium has been a staple of practical jokes among the chemistry cognoscenti ever since, a definite step up from Bunsen-burner humor. One popular trick, since gallium molds easily and looks like aluminium, is to fashion gallium spoons, serve them with tea, and watch as your guests recoil when their Earl Grey "eats" their utensils.*

Lecoq de Boisbaudran reported his findings in scientific

journals, rightfully proud of his capricious metal. Gallium was the first new element discovered since Mendeleev's 1869 table, and when the theorist Mendeleev read about Lecoq de Boisbaudran's work, he tried to cut in line and claim credit for gallium based on his prediction of eka-aluminium. Lecoq de Boisbaudran responded tersely that, no, he had done the real work. Mendeleev demurred, and the Frenchman and Russian began debating the matter in scientific journals, like a serialized novel with different characters narrating each chapter. Before long, the discussion turned acrimonious. Annoyed at Mendeleev's crowing, Lecoq de Boisbaudran claimed an obscure Frenchman had developed the periodic table before Mendeleev and that the Russian had usurped this man's ideas—a scientific sin second only to forging data. (Mendeleev was never so good about sharing credit. Meyer, in contrast, cited Mendeleev's table in his own work in the 1870s, which may have made it seem to later generations that Meyer's work was derivative.)

For his part, Mendeleev scanned Lecoq de Boisbaudran's data on gallium and told the experimentalist, with no justification, that he must have measured something wrong, because the density and weight of gallium differed from Mendeleev's predictions. This betrays a flabbergasting amount of gall, but as science philosopher-historian Eric Scerri put it, Mendeleev always "was willing to bend nature to fit his grand philosophical scheme." The only difference between Mendeleev and crackpottery is that Mendeleev was right: Lecoq de Boisbaudran soon retracted his data and published results that corroborated Mendeleev's predictions. According to Scerri, "The scientific world was astounded to note that Mendeleev, the theorist, had seen the properties of a new element more clearly than the chemist who had discovered it." A literature teacher once told me that what makes a story great—and the construction of the periodic table is a great story—is a climax that's "surprising yet inevitable." I suspect that upon discovering his grand scheme of the periodic table, Mendeleev felt astonished—yet also convinced of its truth because of its elegant, inescapable simplicity. No wonder he sometimes grew intoxicated at the power he felt.

Leaving aside scientific machismo, the real debate here centered on theory versus experiment. Had theory tuned Lecoq de Boisbaudran's senses to help him see something new? Or had experiment provided the real evidence, and Mendeleev's theory just happened to fit? Mendeleev might as well have predicted cheese on Mars before Lecoq de Boisbaudran found evidence for his table in gallium. Then again, the Frenchman had to retract his data and issue new results that supported what Mendeleev had predicted. Although Lecoq de Boisbaudran denied he had ever seen Mendeleev's table, it's possible he had heard of others or that the tables had gotten the scientific community talking and had indirectly primed scientists to keep an eye peeled for new elements. As no less a genius than Albert Einstein once said, "It is theory that decides what we can observe."

In the end, it's probably impossible to tease out whether the heads or tails of science, the theory or the experiment, has done more to push science ahead. That's especially true when you consider that Mendeleev made many wrong predictions. He was lucky, really, that a good scientist like Lecoq de Boisbaudran discovered eka-aluminium first. If someone had poked around for one of his mistakes—Mendeleev predicted there were many elements before hydrogen and swore the sun's halo contained a unique element called coronium—the Russian might have died in obscurity. But just as people forgave ancient astrologers who spun false, even contradictory, horoscopes and fixated instead on the one brilliant comet they predicted exactly, people tend to remember only Mendeleev's triumphs. Moreover, when simplifying history it's tempting to give Mendeleev, as well as Meyer and others, too much credit. They did the important work in building the trellis on which to nail the elements; but by 1869, only two-thirds of all elements had been discovered, and for years some of them sat in the wrong columns and rows on even the best tables.

Loads of work separates a modern textbook from Mendeleev, especially regarding the mess of elements now quarantined at the bottom of the table, the lanthanides. The lanthanides start with lanthanum, element fifty-seven, and their proper home on the table baffled and bedeviled chemists well into the twentieth century. Their buried electrons cause the lanthanides to clump together in frustrating ways; sorting them out was like unknotting kudzu or ivy. Spectroscopy also stumbled with lanthanides, since even if scientists detected dozens of new bands of color, they had no idea how many new elements that translated to. Even Mendeleev, who wasn't shy about predictions, decided the lanthanides were too vexed to make guesses about. Few elements beyond cerium, the second lanthanide, were known in 1869. But instead of chiseling in more "ekas," Mendeleev admitted his helplessness. After cerium, he dotted his table with row after row of frustrating blanks. And later, while filling in new lanthanides after cerium, he often bungled their placement, partly because many "new" elements turned out to be combinations of known ones. It's as if cerium was the edge of the known world to Mendeleev's circle, just like Gibraltar was to ancient mariners, and after cerium they risked falling into a whirlpool or draining off the edge of the earth.

In truth, Mendeleev could have resolved all his frustrations had he traveled just a few hundred miles west from

			K = 39	Rb = 85	Cs = 133	—	—
			Ca = 40	Sr = 87	Ba = 137	—	—
			—	?Yt = 88?	?Di = 138?	Er = 178?	—
			Ti = 48?	Zr = 90	Co = 140?	?La = 180?	Tb = 231
			V = 51	Nb = 94	—	Ta = 182	
			Cr = 52	Mo = 96	—	W = 184	U = 240
			Mn = 55	—	—	—	
			Fe = 56	Ru = 104	—	Os = 195?	
Typische Elemente			Co = 59	Rh = 104	—	Ir = 197	
			Ni = 59	Pd = 106	—	Pt = 198?	
H = 1	Li = 7	Na = 23	Cu = 63	Ag = 108	—	A·. = 199?	—
	Be = 9,4	Mg = 24	Zn = 65	Cd = 112	—	Hg = 200	—
	B = 11	Al = 27,3	—	In = 113	—	Tl = 204	—
	C = 12	Si = 28	—	Sn = 118	—	Pb = 207	—
	N = 14	P = 31	As = 75	Sb = 122	—	Bi = 208	—
	O = 16	S = 32	Se = 78	Te = 125?	—	—	—
	F = 19	Cl = 35,5	Br = 80	J = 127	—	—	—

An early (sideways) periodic table produced by Dmitri Mendeleev in 1869. The huge gap after cerium (Ce) shows how little Mendeleev and his contemporaries knew about the convoluted chemistry of the rare earth metals.

St. Petersburg. There, in Sweden, near where cerium was first discovered, he would have come across a nondescript porcelain mine in a hamlet with the funny name of Ytterby.

In 1701, a braggadocian teenager named Johann Friedrich Böttger, ecstatic at the crowd he'd rallied with a few white lies, pulled out two silver coins for a magic show. After he waved his hands and performed chemical voodoo on them, the silver pieces "disappeared," and a single gold piece materialized in their place. It was the most convincing display of alchemy the locals had ever seen. Böttger thought his reputation was set, and unfortunately it was.

Rumors about Böttger inevitably reached the king of Poland, Augustus the Strong, who arrested the young alchemist and locked him, Rumpelstiltskin-like, in a castle to spin gold for the king's realm. Obviously, Böttger couldn't deliver on this demand, and after a few futile experiments, this harmless liar, still quite young, found himself a candidate for hanging. Desperate to save his neck, Böttger begged the king to spare him. Although he'd failed with alchemy, he claimed he knew how to make porcelain.

At the time, this claim was scarcely more credible. Ever since Marco Polo had returned from China at the end of the thirteenth century, the European gentry had obsessed over white Chinese porcelain, which was hard enough to resist scratching with a nail file yet miraculously translucent like an eggshell. Empires were judged by their tea sets, and wild rumors spread about porcelain's power. One rumor held that you couldn't get poisoned while drinking from a porcelain cup. Another claimed the Chinese were so fabulously wealthy in porcelain that they had erected a nine-story tower of it, just to show off. (That one turned out to be true.) For centuries, powerful Europeans, like the Medici in Florence, had sponsored porcelain research but had succeeded in producing only C-minus knockoffs.

Luckily for Böttger, King Augustus had a capable man working on porcelain, Ehrenfried Walter von Tschirnhaus. Tschirnhaus, whose previous job was to sample the Polish soil to figure out where to dig for crown jewels, had just invented a special oven that reached 3,000°F. This allowed him to melt down porcelain to analyze it, and when the king ordered the clever Böttger to become Tschirnhaus's assistant, the research took off. The duo discovered that the secret ingredients in Chinese porcelain were a white clay called kaolin and a feldspar rock that fuses into glass at high temperatures. Just as crucially, they figured out that, unlike with most crockery, they had to cook the porcelain glaze and clay simultaneously, not in separate steps. It's this high-heat fusion of glaze and clay that gives true porcelain its lucidity and toughness. After perfecting the process, they returned, relieved, to show their liege. Augustus thanked them profusely, dreaming that porcelain would immediately make him, at least socially, the most influential monarch in Europe. And after such a breakthrough, Böttger reasonably expected his freedom. Unfortunately, the king decided he was now too valuable to release and locked him up under tighter security.

Inevitably, the secret of porcelain leaked, and Böttger and Tschirnhaus's recipe spread throughout Europe. With the basic chemistry in place, craftsmen tinkered with and improved the process over the next half century. Soon, wherever people found feldspar, they mined it, including in frosty Scandinavia, where porcelain stoves were prized because they reached higher temperatures and held heat longer than iron-belly stoves. To feed the burgeoning industry in Europe, a feldspar mine opened a dozen miles from Stockholm, on the isle of Ytterby, in 1780.

Ytterby, pronounced "itt-er-bee" and meaning "outer village," looks exactly like you'd hope a coastal village in Sweden would, with red-roofed houses right on the water, big white shutters, and lots of fir trees in roomy yards. People travel around the archipelago in ferries. Streets are named for minerals and elements.*

The Ytterby quarry was scooped from the top of a hill in the southeast corner of the island, and it supplied fine raw ore for porcelain and other purposes. More intriguingly for scientists, its rocks also produced exotic pigments and colored glazes when processed. Nowadays, we know that bright colors are dead giveaways of lanthanides, and the mine in Ytterby was unusually rich in them for a few geological reasons. The earth's elements were once mixed uniformly in the crust, as if someone had dumped a whole rack of spices into a bowl and stirred it. But metal atoms, especially lanthanides, tend to move in herds, and as the molten earth churned, they clumped together. Pockets of lanthanides happened to end up near—actually beneath—Sweden. And because Scandinavia lies near a fault line, tectonic plate action in the remote past plowed the lanthanide-rich rocks up from deep underground, a process aided by Bunsen's beloved hydrothermal vents. Finally, during the last Ice Age, extensive Scandinavian glaciers shaved off the surface of the land. This final geological event exposed the lanthanide-rich rock for easy mining near Ytterby.

But if Ytterby had the proper economic conditions to make mining profitable and the proper geology to make it scientifically worthwhile, it still needed the proper social climate. Scandinavia had barely evolved beyond a Viking mentality by the late 1600s, a century during which even its universities held witch hunts (and sorcerer hunts, for male witches) on a scale that would have embarrassed Salem. But in the 1700s, after Sweden conquered the peninsulas politically and the Swedish Enlightenment conquered it culturally, Scandinavians embraced rationalism en masse. Great scientists started popping up all out of proportion to the region's small population. This included Johan Gadolin, born in 1760, a chemist in a line of scientific-minded academics. (His father occupied a joint professorship in physics and theology, while his grandfather held the even more unlikely posts of physics professor and bishop.)

After extensive travel in Europe as a young man—including in England, where he befriended and toured the clay mines of porcelain maker Josiah Wedgwood—Gadolin settled down in Turku, in what is now Finland, across the Baltic Sea from Stockholm. There he earned a reputation as a geochemist. Amateur geologists began shipping unusual rocks from Ytterby to him to get his opinion, and little by little, through Gadolin's publications, the scientific world began to hear about this remarkable little quarry.

Although he didn't have the chemical tools (or chemical theory) to tweeze out all fourteen lanthanides, Gadolin made significant progress in isolating clusters of them. He made element hunting a pastime, even an avocation, and when, in Mendeleev's old age, chemists with better tools revisited Gadolin's work on the Ytterby rocks, new elements started to fall out like loose change. Gadolin had started a trend by naming one supposed element yttria, and in homage to all the elements' common origin, chemists began to immortalize Ytterby on the periodic table. More elements (seven) trace their lineage back to Ytterby than to any other person, place, or thing. It was the inspiration for ytterbium, yttrium, terbium, and erbium. For the other three unnamed elements, before running out of letters ("rbium" doesn't quite look right), chemists adopted holmium, after Stockholm; thulium, after the mythic name for Scandinavia; and, at Lecoq de Boisbaudran's insistence, Gadolin's namesake, gadolinium.

Overall, of the seven elements discovered in Ytterby, six were Mendeleev's missing lanthanides. History might have been very different—Mendeleev reworked his table incessantly and might have filled in the entire lower realm of the table after cerium by himself—if only he'd made the trip west, across the Gulf of Finland and the Baltic Sea, to this Galápagos Island of the periodic table.

Part II

MAKING ATOMS,
BREAKING ATOMS

4

Where Atoms Come From:
"We Are All Star Stuff"

Where do elements come from? The commonsense view that dominated science for centuries was that they don't come from anywhere. There was a lot of metaphysical jousting over who (or Who) might have created the cosmos and why, but the consensus was that the lifetime of every element coincides with the lifetime of the universe. They're neither created nor destroyed: elements just are. Later theories, such as the 1930s big bang theory, folded this view into their fabric. Since the pinprick that existed back then, fourteen billion years ago, contained all the matter in the universe, everything around us must have been ejected from that speck. Not shaped like diamond tiaras and tin cans and aluminium foil quite yet, but the same basic stuff. (One scientist calculated that it took the big bang ten minutes to create all known matter, then quipped, "The elements were cooked in less time than it takes to cook a dish of duck and roast potatoes.") Again, it's a commonsense view—a stable astrohistory of the elements.

That theory began to fray over the next few decades. German and American scientists had proved by 1939* that the sun and other stars heated themselves by fusing hydrogen together to form helium, a process that releases an outsized amount of energy compared to the atoms' tiny size. Some scientists said, Fine, the population of hydrogen and helium may change, but only slightly, and there's no evidence the populations of other elements change at all. But as telescopes kept improving, more head-scratchers emerged. In theory, the big bang should have ejected elements uniformly in all directions. Yet data proved that most young stars contain only hydrogen and helium, while older stars stew with dozens of elements. Plus, extremely unstable elements such as technetium, which doesn't exist on earth, *do* exist in certain classes of "chemically peculiar stars."* Something must be forging those elements anew every day.

In the mid-1950s, a handful of perceptive astronomers realized that stars themselves are heavenly Vulcans. Though not alone, Geoffrey Burbidge, Margaret Burbidge, William Fowler, and Fred Hoyle did the most to explain the theory of stellar nucleosynthesis in a famous 1957 paper known simply, to the cognoscenti, as B^2FH. Oddly for a scholarly paper, B^2FH opens with two portentous and contradictory quotes from Shakespeare about whether stars govern the fate of mankind.* It goes on to argue they do. It first suggests the universe was once a primordial slurry of hydrogen, with a smattering of helium and lithium. Eventually, hydrogen clumped together into stars, and the extreme gravitational pressure inside stars began fusing hydrogen into helium, a process that fires every star in the sky. But however important cosmologically, the process is dull scientifically, since all stars do is churn out helium for billions of years. Only when the hydrogen burns up, B^2FH suggests—and here is its real contribution—do things start shaking. Stars that sit bovinely for aeons, chewing hydrogen cud, are transformed more profoundly than any alchemist would have dared dream.

Desperate to maintain high temperatures, stars lacking hydrogen begin to burn and fuse helium in their cores. Sometimes helium atoms stick together completely and form even-numbered elements, and sometimes protons and neutrons spall off to make odd-numbered elements. Pretty soon appreciable amounts of lithium, boron, beryllium, and especially carbon accumulate inside stars (and only inside—the cool outer layer remains mostly hydrogen for a star's lifetime). Unfortunately, burning helium releases less energy than burning hydrogen, so stars run through their helium in, at most, a few hundred million years. Some small stars even "die" at this point, creating molten masses of carbon known as white dwarfs. Heavier stars (eight times or so more massive than the sun) fight on, crushing carbon into six more elements, up to magnesium, which buys them a few hundred years. A few more stars perish then, but the biggest, hottest stars (whose interiors reach five billion degrees) burn those elements, too, over a few million years. B^2FH traces these various fusion reactions and explains the recipe for producing everything up to iron: it's nothing less than evolution for elements. As a result of B^2FH, astronomers today can indiscriminately lump every element between lithium and iron together as stellar "metals," and once they've found iron in a star, they don't bother looking for anything smaller—once iron is spotted, it's safe to assume the rest of the periodic table up to that point is represented.

Common sense suggests that iron atoms soon fuse in the biggest stars, and the resulting atoms fuse, forming every element down to the depths of the periodic table. But again, common sense fails. When you do the math and examine how much energy is produced per atomic union, you find that fusing anything to iron's twenty-six protons *costs* energy. That means post-ferric fusion* does an energy-hungry star no good. Iron is the final peal of a star's natural life.

So where do the heaviest elements, twenty-seven through ninety-two, cobalt through uranium, come from? Ironically, says B²FH, they emerge ready-made from mini–big bangs. After prodigally burning through elements such as magnesium and silicon, extremely massive stars (twelve times the size of the sun) burn down to iron cores in about one earth day. But before perishing, there's an apocalyptic death rattle. Suddenly lacking the energy to, like a hot gas, keep their full volume, burned-out stars implode under their own immense gravity, collapsing thousands of miles in just seconds. In their cores, they even crush protons and electrons together into neutrons, until little but neutrons remains there. Then, rebounding from this collapse, they explode outward. And by explode, I mean *explode*. For one glorious month, a supernova stretches millions of miles and shines brighter than a billion stars. And during a supernova, so many gazillions of particles with so much momentum collide so many times per second that they high-jump over the normal energy barriers and fuse onto iron. Many iron nuclei end up coated in neutrons, some of which decay back into protons and thereby create new elements. Every natural combination of element and isotope spews forth from this particle blizzard.

Hundreds of millions of supernovae have gone through this reincarnation and cataclysmic death cycle in our galaxy alone. One such explosion precipitated our solar system. About 4.6 billion years ago, a supernova sent a sonic boom through a flat cloud of space dust about fifteen billion miles wide, the remains of at least two previous stars. The dust particles commingled with the spume from the supernova, and the whole mess began to swirl in pools and eddies, like the bombarded surface of an immense pond. The dense center of the cloud boiled up into the sun (making it a cannibalized remnant of the earlier stars), and planetary bodies began to aggregate and clump together.

The most impressive planets, the gas giants, formed when a stellar wind—a stream of ejecta from the sun—blew lighter elements outward toward the fringes. Among those giants, the gassiest is Jupiter, which for various reasons is a fantasy camp for elements, where they can live in forms never imagined on earth.

Since ancient times, legends about brilliant Venus, ringed Saturn, and Martian-laden Mars have pinged the human imagination. Heavenly bodies provided fodder for the naming of many elements as well. Uranus was discovered in 1781 and so excited the scientific community that, despite the fact that it contains basically zero grams of the element, a scientist named uranium after the new planet in 1789. Neptunium and plutonium sprang from this tradition as well. But of all the planets, Jupiter has had the most spectacular run in recent decades. In 1994, the Shoemaker-Levy 9 comet collided with it, the first intergalactic collision humans ever witnessed. It didn't disappoint: twenty-one comet fragments struck home, and fireballs jumped two thousand miles high. This drama aroused the public, too, and NASA scientists were soon fending off some startling questions during open Q & A sessions online. One man asked if the core of Jupiter might be a diamond larger than the entire earth. Someone else asked what on earth Jupiter's giant red spot had to do with "the hyper-dimensional physics [he'd] been hearing about," the kind of physics that would make time travel possible. A few years after Shoemaker-Levy, when Jupiter's gravity bent the spectacular Hale-Bopp comet toward earth, thirty-nine Nike-clad cultists in San Diego committed suicide because they believed that Jupiter had divinely deflected it and that it concealed a UFO that would beam them to a higher spiritual plane.

Now, there's no accounting for strange beliefs. (Despite his credentials, Fred Hoyle of the B²FH cohort didn't believe in either evolution or the big bang, a phrase he coined derisively on a BBC radio show to pooh-pooh the very idea.) But the diamond question in the previous paragraph at least had foundation in fact. A few scientists once seriously argued (or secretly hoped) that Jupiter's immense mass could produce such a huge gem. Some still hold out hope that liquid diamonds and Cadillac-sized solid ones are possible. And if you're looking for truly exotic materials, astronomers believe that Jupiter's erratic magnetic field can be explained only by oceans of black, liquid "metallic hydrogen." Scientists have seen metallic hydrogen on earth only for nanoseconds under the most exhaustively extreme conditions they can produce. Yet many are convinced that Jupiter has dammed up a reservoir of it twenty-seven thousand miles thick.

The reason elements live such strange lives inside Jupiter (and to a lesser extent inside Saturn, the next-largest planet) is that Jupiter is a 'tweener: not a large planet so much as a failed star. Had Jupiter sucked up about ten times more detritus during its formation, it might have graduated to a brown dwarf, a star with just enough brute mass to fuse some atoms together and give off low-watt, brownish light.* Our solar system would have contained two stars, a binary system. (As we'll see, this isn't so crazy.) Jupiter instead cooled down below the threshold for fusion, but it maintained enough heat and mass and pressure to cram atoms very close together, to the point they stop behaving like the atoms we recognize on earth. Inside Jupiter, they enter a limbo of possibility between chemical and nuclear reactions, where planet-sized diamonds and oily hydrogen metal seem plausible.

The weather on Jupiter's surface plays similar tricks with elements. This shouldn't be surprising on a planet that can support the giant red eye—a hurricane three times wider than the earth that hasn't dissipated after centuries of furious storming. The meteorology deep inside Jupiter is possibly even more spectacular. Because the stellar wind blew only the lightest, most common elements as far out as Jupiter, it should have the same basic elemental composition as real stars—90 percent hydrogen, 10 percent helium, and predictable traces of other elements, including neon. But recent satellite observations showed that a quarter of the helium is missing from the outer atmosphere, as is 90 percent of the neon. Not coincidentally, there is an abundance of those elements deeper down. Something apparently had pumped helium and neon from one spot to the other, and scientists soon realized a weather map could tell them what.

In a real star, all the mini–nuclear booms in the core counterbalance the constant inward tug of gravity. In Jupiter, because it lacks a nuclear furnace, little can stop the heavier helium or neon in the outer, gaseous layers from falling inward. About a quarter of the way into Jupiter, those gases draw close to the liquid metallic hydrogen layer, and the intense atmospheric pressure there crushes the dissolved gas atoms together into liquids. They quickly precipitate out.

Now, everyone has seen helium and neon burning bright colors in tubes of glass—so-called neon lights. The friction from skydiving above Jupiter would have excited falling droplets of those elements in the same way, energizing them like meteors. So if big enough droplets fell far enough fast enough, someone floating right near the metallic hydrogen layer inside Jupiter maybe, just maybe, could have looked up into its cream and orange sky and seen the most spectacular light show ever— fireworks lighting up the Jovian night with a trillion streaks of brilliant crimson, what scientists call neon rain.

*   *   *

The history of our solar system's rocky planets (Mercury, Venus, Earth, and Mars) is different, their drama subtler. When the solar system began to coalesce, the gas giants formed first, in as little as a million years, while the heavy elements congregated in a celestial belt roughly centered on the earth's orbit and stayed quiet for millions of years more. When the earth and its neighbors were finally spun into molten globes, those elements were blended more or less uniformly inside them. *Pace* William Blake, you could have scooped up a handful of soil and held the whole universe, the whole periodic table, in your palm. But as the elements churned around, atoms began tagging up with their twins and chemical cousins, and, after billions of passes up and down, healthy-sized deposits of each element formed. Dense iron sank to the core inside each planet, for instance, where it rests today. (Not to be outdone by Jupiter, Mercury's liquid core sometimes releases iron "snowflakes" shaped not like our planet's familiar, water-based hexagons, but like microscopic cubes.*) The earth might have ended up as nothing but huge floes of uranium and aluminium and other elements, except that something else happened: the planet cooled and solidified enough to make churning difficult. So we're left today with clusters of elements, but enough clusters spread far enough apart that—except in a few notorious cases—no one country monopolizes their supply.

Compared to planets around other stars, our system's four rocky planets have different abundances of each type of element. Most solar systems probably formed from supernovae, and each system's exact elemental ratios depend on the supernova energy available beforehand to fuse elements and also what was present (like space dust) to mix with the ejecta. As a result, each solar system has a unique elemental signature. From high school chemistry you probably remember seeing a number below each element on the periodic table to indicate its atomic weight—the number of protons plus the number of neutrons. Carbon weighs 12.011 units, for example. In reality, that's just an average. Most carbon atoms weigh exactly 12 units, and the 0.011 gets tacked on to account for the scattered carbons that weigh 13 or 14 units. In a different galaxy, however, carbon's average could stray slightly higher or lower. Furthermore, supernovae produce many radioactive elements, which start decaying immediately after the explosion. It's highly unlikely two systems would have the same ratio of radioactive to nonradioactive elements unless both systems were born at once.

Given the variability among solar systems, and given that their formation took place incomprehensibly long ago, reasonable people might ask how scientists have the foggiest idea how the earth was formed. Basically, scientists analyzed the amount and placement of common and rare elements in the earth's crust and deduced how they could have gotten where they are. For instance, the common elements lead and uranium fixed the birth date of the planet through a series of almost insanely meticulous experiments done by a graduate student in Chicago in the 1950s.

The heaviest elements are radioactive, and almost all—most notably uranium—break down into steady lead. Since Clair Patterson came up professionally after the Manhattan Project, he knew the precise rate at which uranium breaks down. He also knew that three kinds of lead exist on earth. Each type, or isotope, has a different atomic weight—204, 206, or 207. Some lead of all three types has existed since our supernova birth, but some has been created fresh by uranium. The catch is that uranium breaks down into only two of those types, 206 and 207. The amount of 204 is fixed, since no element breaks down into it. The key insight was that the ratio of 206 and 207 to the fixed 204 isotope has increased at a predictable rate, because uranium keeps making more of the former two. If Patterson could figure out how much higher that ratio was now than originally, he could use the rate of uranium decay to extrapolate backward to year zero.

The spit in the punch bowl was that no one was around to record the original lead ratios, so Patterson didn't know when to stop tracing backward. But he found a way around that. Not all the space dust around the earth coagulated into planets, of course. Meteors, asteroids, and comets formed, too. Because they formed from the same dust and have floated around in cryogenic space since then, those objects are preserved hunks of primordial earth. What's more, because iron sits at the apex of the stellar nucleosynthesis pyramid, the universe contains a disproportionate amount. Meteors are solid iron. The good news is that, chemically, iron and uranium don't mix, but iron and lead do, so meteors contain lead in the same original ratios as the earth did, because no uranium was around to add new lead atoms. Patterson excitedly got hold of meteor bits from Canyon Diablo in Arizona and got to work.

Only to be derailed by a bigger, more pervasive problem: industrialization. Humans have used soft, pliable lead since ancient times for projects like municipal water pipes. (Lead's symbol on the periodic table, Pb, derives from the same Latin word that gave us "plumber.") And since the advent of lead paint and leaded "anti-knock" gasoline in the late nineteenth and early twentieth centuries, ambient lead levels had been rising the way carbon dioxide levels are rising now. This pervasiveness ruined Patterson's early attempts to analyze meteors, and he had to devise ever more drastic measures—such as boiling equipment in concentrated sulfuric acid—to keep vaporized human lead out of his pristine space rocks. As he later told an interviewer, "The lead from your hair, when you walk into a super-clean laboratory like mine, will contaminate the whole damn laboratory."

This scrupulousness soon morphed into obsession. When reading the Sunday comics, Patterson began to see Pig-Pen, the dust-choked *Peanuts* character, as a metaphor for humanity, in that Pig-Pen's perpetual cloud was our airborne lead. But Patterson's lead fixation did lead to two important results. First, when he'd cleaned up his lab enough, he came up with what's still the best estimate of the earth's age, 4.55 billion years. Second, his horror over lead contamination turned him into an activist, and he's the largest reason future children will never eat lead paint chips and gas stations no longer bother to advertise "unleaded" on their pumps. Thanks to Patterson's crusade, it's common sense today that lead paint should be banned and cars shouldn't vaporize lead for us to breathe in and get in our hair.

Patterson may have pinned down the earth's origin, but knowing when the earth was formed isn't everything. Venus, Mercury, and Mars were formed simultaneously, but except for superficial details, they barely resemble Earth. To piece together the fine details of our history, scientists had to explore some obscure corridors of the periodic table.

In 1977, a father-son physicist-geologist team, Luis and Walter Alvarez, were studying limestone deposits in Italy from about the time the dinosaurs died out. The layers of limestone looked uniform, but a fine, unaccountable layer of red clay dusted the deposits from around the date of extinction, sixty-five million years ago. Strangely, too, the clay contained six hundred times the normal level of the element iridium. Iridium is a siderophile, or iron-loving, element,* and as a result most of it is tied up in the earth's molten iron core. The only common source of iridium is iron-rich meteors, asteroids, and comets—which got the Alvarezes thinking.

Bodies like the moon bear crater scars from ancient bombardments, and there's no reason to think the earth escaped such bombardments. If a huge something the size of a metropolis struck the earth sixty-five million years ago, it would have sent up a Pig-Pen-esque layer of iridium-rich dust worldwide. This cloud would have blotted out the sun and choked off plant life, which all in all seemed a tidy explanation for why not just dinosaurs but 75 percent of all species and 99 percent of all living beings died out around that time. It took a lot of work to convince some scientists, but the Alvarezes soon determined that the iridium layer extended around the world, and they ruled out the competing possibility that the dust deposits had come from a nearby supernova. When other geologists (working for an oil company) discovered a crater more than one hundred miles wide, twelve miles deep, and sixty-five million years old on the Yucatán Peninsula in Mexico, the asteroid-iridium-extinction theory seemed proved.

Except there was still a tiny doubt, a snag on people's scientific conscience. Maybe the asteroid had blackened the sky and caused acid rain and mile-high tsunamis, but in that case the earth would have settled down within decades at most. The trouble was, according to the fossil record, the dinosaurs died out over hundreds of thousands of years. Many geologists today believe that massive volcanoes, in India, which were coincidentally erupting before and after the Yucatán impact, helped kill off the dinosaurs. And in 1984, some paleontologists began arguing that the dinosaur die-off was part of a larger pattern: every twenty-six million years or so, the earth seems to have undergone mass extinctions. Was it just a coincidence that an asteroid fell when the dinosaurs were due?

Geologists also began unearthing other thin layers of iridium-rich clay—which seemed to coincide geologically with other extinctions. Following the Alvarezes' lead, a few people concluded that asteroids or comets had caused all the major wipeouts in the earth's history. Luis Alvarez, the father in the father-son team, found this idea dubious, especially since no one could explain the most important and most radically implausible part of the theory—the cause of the consistency. Fittingly, what reversed Alvarez's opinion was another nondescript element, rhenium.

As Alvarez's colleague Richard Muller recalled in the book *Nemesis*, Alvarez burst into Muller's office one day in the 1980s waving a "ridiculous" and speculative paper on periodic extinctions that he was supposed to peer-review. Alvarez already appeared to be in a froth, but Muller decided to goad Alvarez anyway. The two began arguing like spouses, complete with quivering lips. The crux of the matter, as Muller summarized it, was this: "In the vastness of space, even the Earth is a very small target. An asteroid passing close to the sun has only slightly better than one chance in a billion of hitting our planet. The impacts that do occur should be randomly spaced, not evenly strung out in time. What could make them hit on a regular schedule?"

Even though he had no clue, Muller defended the *possibility* that something could cause periodic bombardments. Finally, Alvarez had had enough with conjectures and called Muller out, demanding to know what that something was. Muller, in what he described as an adrenaline-fueled moment of improvised genius, reached down and blurted out that maybe the sun had a roaming companion star, around which the earth circled too slowly for us to notice—and, and, and whose gravity yanked asteroids toward the earth as it approached us. *Take that!*

Muller might have meant the companion star, later dubbed Nemesis* (after the Greek goddess of retribution), only half-seriously. Nevertheless, the idea stopped Alvarez short, because it explained a tantalizing detail about rhenium. Remember that all solar systems have a signature, a unique ratio of isotopes. Traces of rhenium had been found blended in the layers of iridium clay, and based on the ratio of two types of rhenium (one radioactive, one not), Alvarez knew that any purported asteroids of doom had to have come from our home solar system, since that ratio was the same as on earth. If Nemesis really did swing on by every twenty-six million years and sling space rocks at us, those rocks would also have the same ratio of rhenium. Best of all, Nemesis could explain why the dinosaurs died out so slowly. The Mexican crater might have been only the biggest blow in a pummeling that lasted many thousands of years, as long as Nemesis was in the neighborhood. It might not have been one massive wound but thousands or millions of small stings that ended the famous age of the terrible lizards.

That day in Muller's office, Alvarez's temper—easy come, easy go—evaporated as soon as he realized that periodic asteroids were at least possible. Satisfied, he left Muller alone. But Muller couldn't let go of the serendipitous idea, and the more he pondered it, the more he grew convinced. Why couldn't Nemesis exist? He started talking to other astronomers and publishing papers on Nemesis. He gathered evidence and momentum and wrote his book. For a few glorious years in the mid-1980s, it seemed that even if Jupiter didn't have enough mass to fire itself up as a star, maybe the sun had a celestial companion after all.

Unfortunately, the noncircumstantial evidence for Nemesis was never strong, and it soon looked even scantier. If the original single-impact theory had drawn fire from critics, the Nemesis theory had them lined up to volley bullets like redcoats in the Revolutionary War. It's unlikely that astronomers had simply missed a heavenly body in thousands of years of scanning the sky, even if Nemesis was at its farthest point away. Especially not since the nearest known star, Alpha Centauri, is four light-years away, while Nemesis would have had to inch within half a light-year to inflict its retribution. There are holdouts and romantics still scouring our cosmic zip code for Nemesis, but every year without a sighting makes Nemesis more unlikely.

Still, never underestimate the power of getting people thinking. Given three facts—the seemingly regular extinctions; the iridium, which implies impacts; and the rhenium, which implies projectiles from our solar system—scientists felt they were onto something, even if Nemesis wasn't the mechanism. They hunted for other cycles that could wreak havoc, and they soon found a candidate in the motion of the sun.

Many people assume that the Copernican revolution tacked the sun to a fixed spot in space-time, but really, the sun is dragged along in the tides of our local spiral galaxy and bobs up and down like a carousel as it drifts.* Some scientists think this bobbing brings it close enough to tug on an enormous drifting cloud of comets and space debris that surround our solar system, the Oort cloud. Oort cloud objects all originated with our supernova birth, and whenever the sun climbs to a peak or sinks to a trough every twenty-some million years, it might attract small, unfriendly bodies and send them screaming toward earth. Most would get deflected by the gravity of the sun (or Jupiter, which took the Shoemaker-Levy bullet for us), but enough would slip through that earth would get pummeled. This theory is far from proved, but if it ever is, we're on one long, deadly carousel ride through the universe. At least we can thank iridium and rhenium for letting us know that, perhaps soon, we'd better duck.

In one sense, the periodic table is actually irrelevant to

studying the astrohistory of the elements. Every star consists of virtually nothing but hydrogen and helium, as do gas giant planets. But however important cosmologically, the hydrogen-helium cycle doesn't exactly fire the imagination. To extract the most interesting details of existence, such as supernova explosions and carboniferous life, we need the periodic table. As philosopher-historian Eric Scerri writes, "All the elements other than hydrogen and helium make up just 0.04 percent of the universe. Seen from this perspective, the periodic system appears to be rather insignificant. But the fact remains that we live on the earth...where the relative abundance of elements is quite different."

True enough, though the late astrophysicist Carl Sagan said it more poetically. Without the nuclear furnaces described in B^2FH to forge elements like carbon, oxygen, and nitrogen, and without supernova explosions to seed hospitable places like earth, life could never form. As Sagan affectionately put it, "We are all star stuff."

Unfortunately, one sad truth of astrohistory is that Sagan's "star stuff" didn't grace every part of our planet equally. Despite supernovae exploding elements in all directions, and despite the best efforts of the churning, molten earth, some lands ended up with higher concentrations of rare minerals. Sometimes, as in Ytterby, Sweden, this inspires scientific genius. Too often it inspires greed and rapaciousness—especially when those obscure elements find use in commerce, war, or, worst of all, both at once.

5

Elements in Times of War

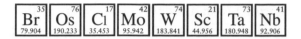

Like other staples of modern society—democracy, philosophy, drama—we can trace chemical warfare back to ancient Greece. The city-state of Sparta, laying siege to Athens in the 400s BC, decided to gas its stubborn rival into submission with the most advanced chemical technology of the time—smoke. Tight-lipped Spartans crept up to Athens with noxious bundles of wood, pitch, and stinky sulfur; lit them; and crouched outside the city walls, waiting for coughing Athenians to flee, leaving their homes unguarded. Though as brilliant an innovation as the Trojan horse, the tactic failed. The fumes billowed through Athens, but the city survived the stink bomb and went on to win the war.*

That failure proved a harbinger. Chemical warfare progressed fitfully, if at all, for the next twenty-four hundred years and remained far inferior to, say, pouring boiling oil on attackers. Up until World War I, gas had little strategic value. Not that countries didn't recognize the threat. All the scientifically advanced nations in the world, save one holdout, signed the Hague Convention of 1899 to ban chemical-based weapons in war. But the holdout, the United States, had a point: banning gases that at the time were hardly more powerful than pepper spray seemed hypocritical if countries were all too happy to mow down eighteen-year-olds with machine guns and sink warships with torpedoes and let sailors drown in the dark sea. The other countries scoffed at U.S. cynicism, ostentatiously signed the Hague pact, and promptly broke their word.

Early, secret work on chemical agents centered on bromine, an energetic grenade of an element. Like other halogens, bromine has seven electrons in its outer energy level but desperately wants eight. Bromine figures that the end justifies the means and shreds the weaker elements in cells, such as carbon, to get its electron fix. Bromine especially irritates the eyes and nose, and by 1910 military chemists had developed bromine-based lacrimators so potent they could incapacitate even a grown man with hot, searing tears.

Having no reason to refrain from using lacrimators on its own citizens (the Hague pact concerned only warfare), the French government collared a ring of Parisian bank robbers with ethyl bromoacetate in 1912. Word of this event quickly spread to France's neighbors, who were right to worry. When war broke out in August 1914, the French immediately lobbed bromine shells at advancing German troops. But even Sparta two millennia before had done a better job. The shells landed on a windy plain, and the gas had little effect, blowing away before the Germans realized they'd been "attacked." However, it's more accurate to say the shells had little *immediate* effect, since hysterical rumors of the gas tore through newspapers on both sides of the conflict. The Germans fanned the flames— blaming an unlucky case of carbon monoxide poisoning in their barracks on secret French asphyxiants, for instance—to justify their own chemical warfare program.

Thanks to one man, a bald, mustached chemist who wore a pince-nez, the German gas research units soon outpaced the rest of the world's. Fritz Haber had one of the great minds in history for chemistry, and he became one of the most famous scientists in the world around 1900 when he figured out how to convert the commonest of chemicals—the nitrogen in air—into an industrial product. Although nitrogen gas can suffocate unsuspecting people, it's usually benign. In fact, it's benign almost to the point of uselessness. The one important thing nitrogen does is replenish soil: it's as crucial to plants as vitamin C is to humans. (When pitcher plants and Venus flytraps trap insects, it's the bugs' nitrogen they're after.) But even though nitrogen makes up 80 percent of air—four of every five molecules we breathe—it's surprisingly bad at topping off soil because it rarely reacts with anything and never becomes "fixed" in the soil. That combination of plentitude, ineptitude, and importance proved a natural target for ambitious chemists.

There are many steps in the process Haber invented to "capture" nitrogen, and many chemicals appear and disappear. But basically, Haber heated nitrogen to hundreds of degrees, injected some hydrogen gas, turned up the pressure to hundreds of times greater than normal air pressure, added some crucial osmium as a catalyst, and voilà: common air transmuted into ammonia, NH_3, the precursor of all fertilizers. With cheap industrial fertilizers now available, farmers no longer were limited to compost piles or dung to nourish their soil. Even by the time World War I broke out, Haber had likely saved millions from Malthusian starvation, and we can still thank him for feeding most of the world's 6.7 billion people today.*

What's lost in that summary is that Haber cared little about fertilizers, despite what he sometimes said to the contrary. He actually pursued cheap ammonia to help Germany build nitrogen explosives—the sort of fertilizer-distilled bombs that

Timothy McVeigh used to blow a hole in an Oklahoma City courthouse in 1995. It's a sad truth that men like Haber pop up frequently throughout history—petty Fausts who twist scientific innovations into efficient killing devices. Haber's story is darker because he was so skilled. After World War I broke out, German military leaders, hoping to break the trench stalemate ruining their economy, recruited Haber for their gas warfare division. Though set to make a fortune from government contracts based on his ammonia patents, Haber couldn't throw away his other projects fast enough. The division was soon referred to as "the Haber office," and the military even promoted Haber, a forty-six-year-old Jewish convert to Lutheranism (it helped his career), to captain, which made him childishly proud.

His family was less impressed. Haber's über alles stance chilled his personal relationships, especially with the one person who might have redeemed him, his wife, Clara Immerwahr. She also exuded genius, becoming the first woman to earn a Ph.D. from the prestigious university in Haber's hometown, Breslau (now Wrocław). But unlike Marie Curie, a contemporary of hers, Immerwahr never came into her own, because instead of marrying an open-minded man like Pierre Curie, she married Haber. On its face, the marriage was not a poor choice for someone with scientific ambitions, but whatever Haber's chemical brilliance, he was a flawed human being. Immerwahr, as one historian puts it, "was never out of apron," and she once rued to a friend about "Fritz's way of putting himself first in our home and marriage, so that a less ruthlessly assertive personality was simply destroyed." She supported Haber by translating manuscripts into English and providing technical support on the nitrogen projects, but she refused to help on the bromine gas work.

Haber barely noticed. Dozens of other young chemists

had volunteered, since Germany had fallen behind the hated French in chemical warfare, and by early 1915 the Germans had an answer to the French lacrimators. Perversely, however, the Germans tested their shells on the British army, which had no gas. Fortunately, as in the first French gas attack, the wind dispersed the gas, and the British targets—bored out of their skulls in a nearby trench—had no idea they'd been attacked.

Undeterred, the German military wanted to devote even more resources to chemical warfare. But there was a problem—that pesky Hague pact, which political leaders didn't want to break (again) publicly. The solution was to interpret the pact in an ultraconscientious yet ultimately bogus way. In signing it, Germany had agreed to "abstain from the use of projectiles, the sole object of which is the diffusion of asphyxiating or deleterious gases." So to the Germans' sophisticated, legalistic reading, the pact had no jurisdiction over shells that delivered shrapnel *and* gas. It took some cunning engineering—the sloshing liquid bromine, which evaporated into gas on impact, wreaked havoc with the shells' trajectory—but Germany's military-industrial-scientific complex prevailed, and a 15 cm shell filled with xylyl bromide, a caustic tearjerker, was ready by late 1915. The Germans called it *weisskreuz*, or "white cross." Again leaving the French alone, Germany swung its mobile gas units east, to shell the Russian army with eighteen thousand *weisskreuze*. If anything, this attempt was more of a debacle than the first. The temperature in Russia was so cold the xylyl bromide froze solid.

Surveying the poor field results, Haber ditched bromine and redirected his efforts to its chemical cousin, chlorine. Chlorine sits above bromine on the periodic table and is even nastier to breathe. It's more aggressive in attacking other elements for one more electron, and because chlorine is smaller—each atom weighs less than half of a bromine atom—chlorine can attack the body's cells much more nimbly. Chlorine turns victims' skin yellow, green, and black, and glasses over their eyes with cataracts. They actually die of drowning, from the fluid buildup in their lungs. If bromine gas is a phalanx of foot soldiers clashing with the mucous membranes, chlorine is a blitzkrieg tank rushing by the body's defenses to tear apart the sinuses and lungs.

Because of Haber, the buffoonery of bromine warfare gave way to the ruthless chlorine phase history books memorialize today. Enemy soldiers soon had to fear the chlorine-based *grunkreuz*, or "green cross"; the *blaukreuz*, or "blue cross"; and the nightmarish blister agent *gelbkreuz*, or "yellow cross," otherwise known as mustard gas. Not content with scientific contributions, Haber directed with enthusiasm the first successful gas attack in history, which left five thousand bewildered Frenchmen burned and scarred in a muddy trench near Ypres. In his spare time, Haber also coined a grotesque biological law, Haber's Rule, to quantify the relationship between gas concentration, exposure time, and death rate—which must have required a depressing amount of data to produce.

Horrified by the gas projects, Clara confronted Fritz early on and demanded he cease. As usual, Fritz listened to her not at all. In fact, although he wept, quite unironically, when colleagues died during an accident in the research branch of the Haber office, after he returned from Ypres he threw a dinner party to celebrate his new weapons. Worse, Clara found out he'd come home just for the night, a stopover on his way to direct more attacks on the eastern front. Husband and wife quarreled violently, and later that night Clara walked into the family garden with Fritz's army pistol and shot herself in the chest. Though no doubt upset, Fritz did not let this inconvenience him. Without staying to make funeral arrangements, he left as planned the next morning.

Despite having the incomparable advantage of Haber, Germany ultimately lost the war to end all wars and was universally denounced as a scoundrel nation. The international reaction to Haber himself was more complicated. In 1919, before the dust (or gas) of World War I had settled, Haber won the vacant 1918 Nobel Prize in chemistry (the Nobels were suspended during the war) for his process to produce ammonia from nitrogen, even though his fertilizers hadn't protected thousands of Germans from famine during the war. A year later, he was charged with being an international war criminal for prosecuting a campaign of chemical warfare that had maimed hundreds of thousands of people and terrorized millions more—a contradictory, almost self-canceling legacy.

Things got worse. Humiliated at the huge reparations Germany had to pay to the Allies, Haber spent six futile years trying to extract dissolved gold from the oceans, so that he could pay the reparations himself. Other projects sputtered along just as uselessly, and the only thing Haber gained attention for during those years (besides trying to sell himself as a gas warfare adviser to the Soviet Union) was an insecticide. Haber had invented Zyklon A before the war, and a German chemical company tinkered with his formula after the war to produce an efficient second generation of the gas. Eventually, a new regime with a short memory took over Germany, and the Nazis soon exiled Haber for his Jewish roots. He died in 1934 while traveling to England to seek refuge. Meanwhile, work on the insecticide continued. And within years the Nazis were gassing millions of Jews, including relatives of Haber, with that second-generation gas—Zyklon B.

*   *   *

In addition to Haber's being a Jew, Germany excommunicated him because he had become passé. In parallel with its gas warfare investment, the German military had begun to exploit a different pocket of the periodic table during World War I, and it eventually decided that bludgeoning enemy combatants with two metals, molybdenum and tungsten, made more sense than scalding them with chlorine and bromine gas. Once again, then, warfare turned on simple, basic periodic table chemistry. Tungsten would go on to become the "it" metal of the Second World War, but in some ways molybdenum's story is more interesting. Almost no one knows it, but the most remote battle of World War I took place not in Siberia or against Lawrence of Arabia on the Sahara sands, but at a molybdenum mine in the Rocky Mountains of Colorado.

After its gas, Germany's most feared weapons during the war were its Big Berthas, a suite of superheavy siege guns that battered soldiers' psyches as brutally as they did the trenches of France and Belgium. The first Berthas, at forty-three tons, had to be transported in pieces by tractors to a launchpad and took two hundred men six hours to assemble. The payoff was the ability to hurl a 16-inch, 2,200-pound shell nine miles in just seconds. Still, a big flaw hobbled the Berthas. Lofting a one-ton mass took whole kegs of gunpowder, which produced massive amounts of heat, which in turn scorched and warped the twenty-foot steel barrels. After a few days of hellish shooting, even if the Germans limited themselves to a few shots per hour, the gun itself was shot to hell.

Never at a loss when providing weaponry for the fatherland, the famous Krupp armament company found a recipe for strengthening steel: spiking it with molybdenum. Molybdenum (pronounced "mo-lib-di-num") could withstand the excessive heat because it melts at 4,750°F, thousands of degrees hotter than iron, the main metal in steel. Molybdenum's atoms are larger than iron's, so they get excited more slowly, and they have 60 percent more electrons, so they absorb more heat and bind together more tightly. Plus, atoms in solids spontaneously and often disastrously rearrange themselves when temperatures change (more on this in chapter 16), which often results in brittle metals that crack and fail. Doping steel with molybdenum gums up the iron atoms, preventing them from sliding around. (The Germans were not the first ones to figure this out. A master sword maker in fourteenth-century Japan sprinkled molybdenum into his steel and produced the island's most coveted samurai swords, whose blades never dulled or cracked. But since this Japanese Vulcan died with his secret, it was lost for five hundred years—proof that superior technology does not always spread and often goes extinct.)

Back in the trenches, the Germans were soon blazing away at the French and British with a second generation of "moly steel" guns. But Germany soon faced another huge Bertha setback—it had no supply of molybdenum and risked running out. In fact, the only known supplier was a bankrupt, nearly abandoned mine on Bartlett Mountain in Colorado.

Before World War I, a local had laid claim to Bartlett upon discovering veins of ore that looked like lead or tin. Those metals would have been worth at least a few cents per pound, but the useless molybdenum he found cost more to mine than it fetched, so he sold his mining rights to one Otis King, a feisty five-foot-five banker from Nebraska. Always enterprising, King adopted a new extraction technique that no one had bothered to invent before and quickly liberated fifty-eight hundred pounds of pure molybdenum—which more or less ruined him. Those nearly three tons exceeded the yearly world demand for molybdenum by 50 percent, which meant King hadn't just flooded the market, he'd drowned it. Noting at least the novelty of King's attempt, the U.S. government mentioned it in a mineralogical bulletin in 1915.

Few noticed the bulletin except for a behemoth international mining company based in Frankfurt, Germany, with a U.S. branch in New York. According to one contemporary account, Metallgesellschaft had smelters, mines, refineries, and other "tentacles" all over the world. As soon as the company directors, who had close ties to Fritz Haber, read about King's molybdenum, they mobilized and ordered their top man in Colorado, Max Schott, to seize Bartlett Mountain.

Schott—a man described as having "eyes penetrating to the point of hypnosis"—sent in claim jumpers to set up stakes and harass King in court, a major drain on an already floundering mine. The more belligerent claim jumpers threatened the wives and children of miners and destroyed their camps during a winter in which the temperature dropped to twenty below. King hired a limping outlaw named Two-Gun Adams for protection, but the German agents got to King anyway, mugging him with knives and pickaxes on a mountain pass and hurling him off a sheer cliff. Only a well-placed snowbank saved his neck. As the self-described "tomboy bride" of one miner put it in her memoirs, the Germans did "everything short of downright slaughter to hinder the work of his company." King's gritty workers took to calling the unpronounceable metal they risked their lives to dig up "Molly be damned."

King had a dim idea what Molly did in Germany, but he was about the only non-German in Europe or North America who did. Not until the British captured German arms in 1916 and reverse-engineered them by melting them down did the Allies discover the *wundermetall*, but the shenanigans in the

Rockies continued. The United States didn't enter World War I until 1917, so it had no special reason to monitor Metallgesellschaft's subsidiary in New York, especially considering its patriotic name, American Metal. It was American Metal that Max Schott's "company" answered to, and when the government began asking questions around 1918, American Metal claimed that it legally owned the mine, since the harried Otis King had sold it to Schott for a paltry $40,000. It also admitted that, um, it just happened to ship all that molybdenum to Germany. The feds quickly froze Metallgesellschaft's U.S. stock and took control of Bartlett Mountain. Sadly, those efforts came too late to disable Germany's Big Berthas. As late as 1918, Germany used moly steel guns to shell Paris from the astonishing distance of seventy-five miles.

The only justice was that Schott's company went bankrupt after the armistice, in March 1919, when molybdenum prices bottomed out. King returned to mining and became a millionaire by persuading Henry Ford to use moly steel in car engines. But Molly's days in warfare were over. By the time World War II rolled around, molybdenum had been superseded in steel production by the element below it on the periodic table, tungsten.

Now if molybdenum is one of the harder elements to pronounce on the periodic table, tungsten has one of the most confounding chemical symbols, a big fat unaccountable W. It stands for *wolfram*, the German name for the metal, and that "wolf" correctly portended the dark role it would play in the war. Nazi Germany coveted tungsten for making machinery and armor-piercing missiles, and its lust for *wolfram* surpassed even its lust for looted gold, which Nazi officials happily bartered for tungsten. And who were the Nazis' trading partners? Not Italy and Japan, the other Axis powers. Nor any of the countries German troops overran, such as Poland or Belgium. It was supposedly neutral Portugal whose tungsten fed the wolfish appetite of the German *kriegwerks*.

Portugal was a hard country to figure at the time. It lent the Allies a vital air base in the Azores, a group of islands in the Atlantic Ocean, and as anyone who's seen *Casablanca* knows, refugees longed to escape to Lisbon, from which they could safely fly to Britain or the United States. However, the dictator of Portugal, Antonio Salazar, tolerated Nazi sympathizers in his government and provided a haven for Axis spies. He also rather two-facedly shipped thousands of tons of tungsten to both sides during the war. Proving his worth as a former professor of economics, Salazar leveraged his country's near monopoly on the metal (90 percent of Europe's supply) into profits 1,000 percent greater than peacetime levels. This might have been defensible had Portugal had long-standing trade relations with Germany and been worried about falling into wartime poverty. But Salazar began selling tungsten to Germany in appreciable quantities only in 1941, apparently on the theory that his country's neutral status allowed him to gouge both sides equally.

The tungsten trade worked like this. Learning its lesson with molybdenum and recognizing the strategic importance of tungsten, Germany had tried to stockpile tungsten before it began erasing boundaries between itself and Poland and France. Tungsten is one of the hardest metals known, and adding it to steel made for excellent drill bits and saw heads. Plus, even modest-sized missiles tipped with tungsten—so-called kinetic energy penetrators—could take down tanks. The reason tungsten proved superior to other steel additives can be read right off the periodic table. Tungsten, situated below molybdenum, has similar properties. But with even more electrons, it doesn't melt until 6,200°F. Plus, as a heavier atom than molybdenum, tungsten provides even better anchors against the iron atoms' slipping around. Remember that the nimble chlorine worked well in gas attacks. Here, in a metal, tungsten's solidity and strength proved attractive.

So attractive that the profligate Nazi regime spent its entire tungsten reserve by 1941, at which point the führer himself got involved. Hitler ordered his ministers to grab as much tungsten as the trains across conquered France could carry. Distressingly, far from there being a black market for this grayish metal, the whole process was entirely transparent, as one historian noted. Tungsten was shipped from Portugal through fascist Spain, another "neutral," and much of the gold the Nazis had seized from Jews—including the gold wrenched out of the teeth of gassed Jews—was laundered by banks in Lisbon and Switzerland, still another country that took no sides. (Fifty years on, a major Lisbon bank still maintained that officials had had no idea that the forty-four tons of gold they had received were dirty, despite the swastikas stamped on many bars.)

Even stalwart Britain couldn't be bothered about the tungsten that was helping to cut down its lads. Prime Minister Winston Churchill privately referred to Portugal's tungsten trade as a "misdemeanor," and lest that remark be misconstrued, he added that Salazar was "quite right" to trade tungsten with Britain's avowed enemies. Once again, however, there was a dissenter. All this naked capitalism, which benefited socialist Germany, caused apoplectic fits in the free-market United States. American officials simply couldn't grasp why Britain didn't order, or outright bully, Portugal to drop its profitable neutrality. Only after prolonged U.S. pressure did Churchill agree to help strong-arm the strongman Salazar.

Until then, Salazar (if we lay aside morality for a moment) had played the Axis and Allies brilliantly with vague promises, secret pacts, and stalling tactics that kept the tungsten trains chugging. He had increased the price of his country's one commodity from $1,100 per ton in 1940 to $20,000 in 1941, and he'd banked $170 million in three frenzied years of speculation. Only after running out of excuses did Salazar institute a full tungsten embargo against the Nazis on June 7, 1944— the day after D-Day, by which point the Allied commanders were too preoccupied (and disgusted) to punish him. I believe it was Rhett Butler in *Gone with the Wind* who said that fortunes can be made only during the building up or tearing down of an empire, and Salazar certainly subscribed to that theory. In the so-called wolfram war, the Portuguese dictator had the last lycanthropic laugh.

Tungsten and molybdenum were only the first hints of a veritable metals revolution that would take place later in the twentieth century. Three of every four elements are metals, but beyond iron, aluminium, and a few others, most did nothing but plug holes in the periodic table before World War II. (Indeed, this book could not have been written forty years ago—there wouldn't have been enough to say.) But since about 1950, every metal has found a niche. Gadolinium is perfect for magnetic resonance imaging (MRI). Neodymium makes unprecedentedly powerful lasers. Scandium, now used as a tungstenlike additive in aluminium baseball bats and bike frames, helped the Soviet Union make lightweight helicopters in the 1980s and purportedly even topped Soviet ICBM missiles stored underground in the Arctic, to help the nukes punch through sheets of ice.

Alas, for all the technological advances made during the metals revolution, some elements continued to abet wars—and not in the remote past, but in the past decade. Fittingly, two of these elements were named after two Greek mythological characters known for great suffering. Niobe earned the ire of the gods by bragging about her seven lovely daughters and seven handsome sons—whom the easily offended Olympians soon slaughtered for her impertinence. Tantalus, Niobe's father, killed his own son and served him at a royal banquet. As punishment, Tantalus had to stand for all eternity up to his neck in a river, with a branch loaded with apples dangling above his nose. Whenever he tried to eat or drink, however, the fruit would be blown away beyond his grasp or the water would recede. Still, while elusiveness and loss tortured Tantalus and Niobe, it is actually a surfeit of their namesake elements that has decimated central Africa.

There's a good chance you have tantalum or niobium in your pocket right now. Like their periodic table neighbors, both are dense, heat-resistant, noncorrosive metals that hold a charge well—qualities that make them vital for compact cell phones. In the mid-1990s cell phone designers started demanding both metals, especially tantalum, from the world's largest supplier, the Democratic Republic of Congo, then called Zaire. Congo sits next to Rwanda in central Africa, and most of us probably remember the Rwandan butchery of the 1990s. But none of us likely remembers the day in 1996 when the ousted Rwandan government of ethnic Hutus spilled into Congo seeking refuge. At the time it seemed just to extend the Rwandan conflict a few miles west, but in retrospect it was a brush fire blown right into a decade of accumulated racial kindling. Eventually, nine countries and two hundred ethnic tribes, each with its own ancient alliances and unsettled grudges, were warring in the dense jungles.

Nonetheless, if only major armies had been involved, the Congo conflict likely would have petered out. Larger than

Alaska and dense as Brazil, Congo is even less accessible than either by roads, meaning it's not ideal for waging a protracted war. Plus, poor villagers can't afford to go off and fight unless there's money at stake. Enter tantalum, niobium, and cellular technology. Now, I don't mean to impute direct blame. Clearly, cell phones didn't cause the war—hatred and grudges did. But just as clearly, the infusion of cash perpetuated the brawl. Congo has 60 percent of the world's supply of the two metals, which blend together in the ground in a mineral called coltan. Once cell phones caught on—sales rose from virtually zero in 1991 to more than a billion by 2001—the West's hunger proved as strong as Tantalus's, and coltan's price grew tenfold. People purchasing ore for cell phone makers didn't ask and didn't care where the coltan came from, and Congolese miners had no idea what the mineral was used for, knowing only that white people paid for it and that they could use the profits to support their favorite militias.

Oddly, tantalum and niobium proved so noxious because coltan was so democratic. Unlike the days when crooked Belgians ran Congo's diamond and gold mines, no conglomerates controlled coltan, and no backhoes and dump trucks were necessary to mine it. Any commoner with a shovel and a good back could dig up whole pounds of the stuff in creek beds (it looks like thick mud). In just hours, a farmer could earn twenty times what his neighbor did all year, and as profits swelled, men abandoned their farms for prospecting. This upset Congo's already shaky food supply, and people began hunting gorillas for meat, virtually wiping them out, as if they were so many buffalo. But gorilla deaths were nothing compared to the human atrocities. It's not a good thing when money pours into a country with no government. A brutal form of capitalism took over in which all things, including lives, were for sale. Huge fenced-in "camps" with enslaved prostitutes sprang up, and innumerable bounties were put out for blood killings. Gruesome stories have circulated about proud victors humiliating their victims' bodies by draping themselves with entrails and dancing in celebration.

The fires burned hottest in Congo between 1998 and 2001, at which point cell phone makers realized they were funding anarchy. To their credit, they began to buy tantalum and niobium from Australia, even though it cost more, and Congo cooled down a bit. Nevertheless, despite an official truce ending the war in 2003, things never really calmed down in the eastern half of the country, near Rwanda. And lately another element, tin, has begun to fund the fighting. In 2006, the European Union outlawed lead solder in consumer goods, and most manufacturers have replaced it with tin—a metal Congo also happens to have in huge supply. Joseph Conrad once called Congo "the vilest scramble for loot that ever disfigured the history of human conscience," and there's little reason to revise that notion today.

Overall, more than five million people have died in Congo since the mid-1990s, making it the biggest waste of life since World War II. The fighting there is proof that in addition to all the uplifting moments the periodic table has inspired, it can also play on humankind's worst, most inhuman instincts.

6

Completing the Table . . .
with a Bang

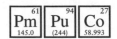

A supernova sowed our solar system with every natural element, and the churning of young molten planets made sure those elements were well blended in the rocky soil. But those processes alone cannot tell us everything about the distribution of elements on earth. Since the supernova, whole species of elements have gone extinct because their nuclei, their cores, were too fragile to survive in nature. This instability shocked scientists and left unaccountable holes in the periodic table — holes that, unlike in Mendeleev's time, scientists just couldn't fill, no matter how hard they searched. They eventually did fill in the table, but only after developing new fields that let them create elements on their own, and only after realizing that the fragility of some elements conceals a bright, shiny danger. The making of atoms and the breaking of atoms proved more intimately bound than anyone dared expect.

The roots of this story go back to the University of Manchester in England just before World War I. Manchester had assembled some brilliant scientists, including lab director Ernest Rutherford. Perhaps the most promising student was

Henry Moseley. The son of a naturalist admired by Charles Darwin, Moseley was drawn instead to the physical sciences. He treated his lab work like a deathbed vigil, staying for fifteen-hour stretches, as if he'd never have time to finish all he wanted to do, and he subsisted on mere fruit salad and cheese. Like many gifted people, Moseley was also a pill, stiff and stuffy, and he expressed open disgust at the "scented dirtiness" of foreigners at Manchester.

But young Moseley's talent excused a lot. Although Rutherford objected to the work as a waste of time, Moseley grew enthusiastic about studying elements by blasting them with electron beams. He enlisted Darwin's grandson, a physicist, as a partner and in 1913 began to systematically probe every discovered element up to gold. As we know today, when a beam of electrons strikes an atom, the beam punches out the atom's own electrons, leaving a hole. Electrons are attracted to an atom's nucleus because electrons and protons have opposite charges, and tearing electrons away from the nucleus is a violent deed. Since nature abhors a vacuum, other electrons rush in to fill the gap, and the crashing about causes them to release high-energy X-rays. Excitingly, Moseley found a mathematical relation between the wavelength of the X-rays, the number of protons an element has in its nucleus, and the element's atomic number (its spot on the periodic table).

Since Mendeleev had published his famous table in 1869, it had undergone a number of changes. Mendeleev had set his first table sideways, until someone showed him the sense in rotating it ninety degrees. Chemists continued to tinker with the table, adding columns and reshuffling elements, over the next forty years. Meanwhile, anomalies had begun to peck at people's confidence that they really understood the table. Most of the elements line up on the table in a cattle call of increasing weight.

According to that criterion, nickel should precede cobalt. Yet to make the elements fit properly—so cobalt sat above cobalt-like elements and nickel above nickel-like elements—chemists had to switch their spots. No one knew why this was necessary, and it was just one of several annoying cases. To get around this problem, scientists invented the atomic number as a placeholder, which just underscored that no one knew what the atomic number actually meant.

Moseley, just twenty-five, solved the riddle by translating the question from chemistry to physics. The crucial thing to realize is that few scientists believed in the atomic nucleus at the time. Rutherford had put forward the idea of a compact, highly positive nucleus just two years earlier, and it remained unproven in 1913, too tentative for scientists to accept. Moseley's work provided the first confirmation. As Niels Bohr, another Rutherford protégé, recalled, "We cannot understand it today, but [the Rutherford work] was not taken seriously.... The great change came from Moseley." That's because Moseley linked an element's place on the table to a physical characteristic, equating the positive nuclear charge with the atomic number. And he did so with an experiment that anyone could repeat. This proved the ordering of elements was not arbitrary but arose from a proper understanding of atomic anatomy. Screwy cases such as cobalt and nickel suddenly made sense, since the lighter nickel had more protons and therefore a higher positive charge and therefore had to come after cobalt. If Mendeleev and others discovered the Rubik's Cube of the elements, Moseley solved it, and after Moseley there was no more need to fudge explanations.

Furthermore, like the spectroscope, Moseley's electron gun helped tidy up the table by sorting through a confusing array of radioactive species and disproving spurious claims for new elements. Moseley also fingered just four remaining holes in the table—elements forty-three, sixty-one, seventy-two, and seventy-five. (The elements heavier than gold were too dear to obtain proper samples to experiment on in 1913. Had Moseley been able to, he would have found gaps at eighty-five, eighty-seven, and ninety-one, too.)

Unfortunately, chemists and physicists mistrusted each other in this era, and some prominent chemists doubted that Moseley had come up with anything as grand as he claimed. Georges Urbain of France challenged the young Turk by bringing him an Ytterby-like blend of ambiguous rare earth elements. Urbain had labored twenty years learning rare earth chemistry, and it had taken him months of tedium to identify the four elements in his sample, so he expected to stymie if not embarrass Moseley. After their initial meeting, Moseley returned to Urbain within an hour with a full and correct list.* The rare earths that had so frustrated Mendeleev were now trivial to sort out.

But they were sorted out by people other than Moseley. Although he pioneered nuclear science, as with Prometheus, the gods punished this young man whose work illuminated the darkness for later generations. When World War I broke out, Moseley enlisted in the king's army (against the army's advice) and saw action in the doomed Gallipoli campaign of 1915. One day the Turkish army rushed the British lines in phalanxes eight deep, and the battle devolved into a street fight with knives, stones, and teeth. Somewhere in that savage scrum, Moseley, age twenty-seven, fell. The futility of that war is best known through the English poets who also died on the battlefield. But one colleague spit that losing Henry Moseley by itself ensured that the war to end all wars would go down as "one of the most hideous and most irreparable crimes in history."*

The best tribute scientists could pay to Moseley was to hunt down all the missing elements he'd pointed out. Indeed, Moseley so inspired element hunters, who suddenly had a clear idea of what to search for, that element safaris became almost too popular. Scuffles soon arose over who'd first bagged hafnium, protactinium, and technetium. Other research groups filled in the gaps at elements eighty-five and eighty-seven in the late 1930s by creating elements in the lab. By 1940, only one natural element, one prize, remained undiscovered—element sixty-one.

Oddly, though, only a few research teams around the world were bothering to look for it. One team, led by an Italian physicist named Emilio Segrè, tried to create an artificial sample and probably succeeded in 1942, but they gave up after a few attempts to isolate it. It wasn't until seven years later that three scientists from Oak Ridge National Laboratory in Tennessee rose at a scientific meeting in Philadelphia and announced that after sifting through some spent uranium ore, they had discovered element sixty-one. After a few hundred years of chemistry, the last hole in the periodic table had been filled.

But the announcement didn't rally much excitement. The trio announced they'd discovered element sixty-one two years before and had sat on the results because they were too preoccupied with their work on uranium—their real work. The press gave the finding correspondingly tepid coverage. In the *New York Times*, the missing link shared a crowded headline with a dubious mining technique that promised a hundred uninterrupted years of oil. *Time* buried the news in its conference wrap-up and pooh-poohed the element as "not good for much."* Then the scientists announced that they planned to name it promethium. Elements discovered earlier in the century had been given boastful or at least explanatory names, but promethium—after the Titan in Greek mythology who stole fire, gave it to humankind, and was tortured by having a vulture dine on his liver—evoked something stern and grim, even guilty.

So what happened between Moseley's time and the discovery of element sixty-one? Why had hunting for elements gone from work so important that a colleague had called Moseley's death an irreparable crime to work worth barely a few lines of newsprint? Sure, promethium was useless, but scientists, of all people, cheer impractical discoveries, and the completion of the periodic table was epochal, the culmination of millions of man-hours. Nor had people simply gotten fatigued with seeking new elements—that pursuit caused sparring between American and Soviet scientists through much of the cold war. Instead, the nature and enormity of nuclear science had changed. People had *seen things*, and a mid-range element like promethium could no longer rouse them like the heavy elements plutonium and uranium, not to mention their famous offspring, the atomic bomb.

One morning in 1939, a young physicist at the University of California at Berkeley settled into a pneumatic barber's chair in the student union for a haircut. Who knows the topic of conversation that day—maybe that son of a bitch Hitler or whether the Yankees would win their fourth straight World Series. Regardless, Luis Alvarez (not yet famous for his dinosaur extinction theory) was chatting and leafing through the *San Francisco Chronicle* when he ran across a wire service item about experiments by Otto Hahn in Germany, on fission—the splitting of the uranium atom. Alvarez halted his barber "mid-snip," as a friend recalled, tore off his smock, and sprinted up the road to his laboratory, where he scooped up a Geiger counter and made a beeline for some irradiated uranium. His hair still only half-cut, he summoned everyone within shouting distance to come see what Hahn had discovered.

Beyond being amusing, Alvarez's dash symbolizes the state of nuclear science at the time. Scientists had been making steady if slow progress in understanding how the cores of atoms work, little snippets of knowledge here and there—and then, with one discovery, they found themselves on a mad tear.

Moseley had given atomic and nuclear science legitimate footing, and loads of talent had poured into those fields in the 1920s. Nevertheless, gains had proved more difficult than expected. Part of the confusion was, indirectly, Moseley's fault. His work had proved that isotopes such as lead-204 and lead-206 could have the same net positive charge yet have different atomic weights. In a world that knew only about protons and electrons, this left scientists floundering with unwieldy ideas about positive protons in the nucleus that gobbled up negative electrons Pac-Man style.* In addition, to comprehend how subatomic particles behave, scientists had to devise a whole new mathematical tool, quantum mechanics, and it took years to figure out how to apply it to even simple, isolated hydrogen atoms.

Meanwhile, scientists were also developing the related field of radioactivity, the study of how nuclei fall apart. Any old atom can shed or steal electrons, but luminaries such as Marie Curie and Ernest Rutherford realized that some rare elements could alter their nuclei, too, by blowing off atomic shrapnel. Rutherford especially helped classify all the shrapnel into just a few common types, which he named using the Greek alphabet, calling them alpha, beta, or gamma decay. Gamma decay is the simplest and deadliest—it occurs when the nucleus emits concentrated X-rays and is today the stuff of nuclear nightmares.

The other types of radioactivity involve the conversion of one element to another, a tantalizing process in the 1920s. But each element goes radioactive in a characteristic way, so the deep, underlying features of alpha and beta decay baffled scientists, who were growing increasingly frustrated about the nature of isotopes as well. The Pac-Man model was failing, and a few daredevils suggested that the only way to deal with the proliferation of new isotopes was to scrap the periodic table.

The giant collective forehead slap—the "Of course!" moment—took place in 1932, when James Chadwick, yet another of Rutherford's students, discovered the neutral neutron, which adds weight without charge. Coupled with Moseley's insights about the atomic number, atoms (at least lone, isolated atoms) suddenly made sense. The neutron meant that lead-204 and lead-206 could still both be lead—could still have the same positive nuclear charge and sit in the same box on the periodic table—even if they had different atomic weights. The nature of radioactivity suddenly made sense, too. Beta decay was understood as the conversion of neutrons to protons or vice versa—and it's because the proton number changes that beta decay converts an atom into a different element. Alpha decay also converts elements and is the most dramatic change on a nuclear level—two neutrons and two protons are shorn away.

Over the next few years, the neutron became more than a theoretical tool. For one thing, it supplied a fantastic way to probe atomic innards, because scientists could shoot a neutron at atoms without it being electrically repulsed, as charged projectiles were. Neutrons also helped scientists induce a new type of radioactivity. Elements, especially lighter elements, try to maintain a rough one-to-one ratio of neutrons to protons. If an atom has too many neutrons, it splits itself, releasing energy and excess neutrons in the process. If nearby atoms absorb those neutrons, they become unstable and spit out more neutrons, a cascade known as a chain reaction. A physicist named Leo Szilard dreamed up the idea of a nuclear chain reaction circa 1933 while standing at a London stoplight one morning. He patented it in 1934 and tried (but failed) to produce a chain reaction in a few light elements as early as 1936.

But notice the dates here. Just as the basic understanding of electrons, protons, and neutrons fell into place, the old-world political order was disintegrating. By the time Alvarez read about uranium fission in his barber's smock, Europe was doomed.

The genteel old world of element hunting died at the same time. With their new model of atomic innards, scientists began to see that the few undiscovered elements on the periodic table were undiscovered because they were intrinsically unstable. Even if they had existed in abundance on the early earth, they had long since disintegrated. This conveniently explained the holes in the periodic table, but the work proved its own undoing. Probing unstable elements soon led scientists to stumble onto nuclear fission and neutron chain reactions. And as soon as they understood that atoms could be split—understood both the scientific and political implications of that fact—collecting new elements for display seemed like an amateur's hobby, like the fusty, shoot-and-stuff biology of the 1800s compared with molecular biology today. Which is why, with a world war and the possibility of atomic bombs staring at them in 1939, no scientists bothered tracking promethium down until a decade later.

No matter how keyed up scientists got about the possibility of fission bombs, however, a lot of work still separated the theory from the reality. It's hard to remember today, but nuclear bombs were considered a long shot at best, especially by military experts. As usual, those military leaders were eager to enlist scientists in World War II, and the scientists dutifully exacerbated the war's gruesomeness through technology such as better steel. But the war would not have ended with two mushroom clouds had the U.S. government, instead of just demanding bigger, faster weapons *now*, summoned the political will to invest billions in a hitherto pure and impractical field: subatomic science. And even then, figuring out how to divide atoms in a controlled manner proved so far beyond the science of the day that the Manhattan Project had to adopt a whole new research strategy to succeed—the Monte Carlo method, which rewired people's conceptions of what "doing science" meant.

As noted, quantum mechanics worked fine for isolated atoms, and by 1940 scientists knew that absorbing a neutron made an atom queasy, which made it explode and possibly release more neutrons. Following the path of one given neutron was easy, no harder than following a caroming billiard ball. But starting a chain reaction required coordinating billions of billions of neutrons, all of them traveling at different speeds in every direction. This made hash of scientists' built-for-one theoretical apparatus. At the same time, uranium and plutonium were expensive and dangerous, so detailed experimental work was out of the question.

Yet Manhattan Project scientists had orders to figure out exactly how much plutonium and uranium they needed to create a bomb: too little and the bomb would fizzle out. Too much and the bomb would blow up just fine, but at the cost of prolonging the war by months, since both elements were monstrously complicated to purify (or in plutonium's case, synthesize, then purify). So, just to get by, some pragmatic scientists decided to abandon both traditional approaches, theory and experiment, and pioneer a third path.

To start, they picked a random speed for a neutron bouncing around in a pile of plutonium (or uranium). They also picked a random direction for it and more random numbers for other parameters, such as the amount of plutonium available, the chance the neutron would escape the plutonium before being absorbed, even the geometry and shape of the plutonium pile. Note that selecting specific numbers meant that scientists were conceding the universality of each calculation, since the results applied to only a few neutrons in one of many designs. Theoretical scientists *hate* giving up universally applicable results, but they had no other choice.

At this point, rooms full of young women with pencils (many of them scientists' wives, who'd been hired to help out because they were crushingly bored in Los Alamos) would get a sheet with the random numbers and begin to calculate (sometimes without knowing what it all meant) how the neutron collided with a plutonium atom; whether it was gobbled up; how many new neutrons if any were released in the process; how many neutrons those in turn released; and so on. Each of the hundreds of women did one narrow calculation in an assembly line, and scientists aggregated the results. Historian George Dyson described this process as building bombs "*numerically*, neutron by neutron, nanosecond by nanosecond ... [a method] of statistical approximation whereby a random sampling of events ... is followed through a series of representative slices in time, answering the otherwise incalculable question of whether a configuration would go thermonuclear."*

Sometimes the theoretical pile did go nuclear, and this was counted as a success. When each calculation was finished, the women would start over with different numbers. Then do it again. And again. And yet again. Rosie the Riveter may have become the iconic symbol of empowered female employment during the war, but the Manhattan Project would have gone nowhere without these women hand-crunching long tables of data. They became known by the neologism "computers."

But why was this approach so different? Basically, scientists equated each computation with an experiment and collected only virtual data for the plutonium and uranium bombs. They abandoned the meticulous and mutually corrective interplay of theory and lab work and adopted methods one historian described unflatteringly as "dislocated . . . a simulated reality that borrowed from both experimental and theoretical domains, fused these borrowings together, and used the resulting amalgam to stake out a netherland at once nowhere and everywhere on the usual methodological map."*

Of course, such calculations were only as good as scientists' initial equations, but here they got lucky. Particles on the quantum level are governed by statistical laws, and quantum mechanics, for all its bizarre, counterintuitive features, is the single most accurate scientific theory ever devised. Plus, the sheer number of calculations scientists pushed through during the Manhattan Project gave them great confidence — confidence that was proved justified after the successful Trinity test in New Mexico in mid-1945. The swift and flawless detonation of a uranium bomb over Hiroshima and a plutonium bomb over Nagasaki a few weeks later also testified to the accuracy of this unconventional, calculation-based approach to science.

After the isolated camaraderie of the Manhattan Project ended, scientists scattered back to their homes to reflect on what they'd done (some proudly, some not). Many gladly forgot about their time served in the calculation wards. Some, though, were riveted by what they'd learned, including one Stanislaw Ulam. Ulam, a Polish refugee who'd passed hours in New Mexico playing card games, was playing solitaire one day in 1946 when he began wondering about the odds of winning any randomly dealt hand. The one thing Ulam loved more than cards was futile calculation, so he began filling pages with probabilistic equations. The problem soon ballooned to such complexity that Ulam smartly gave up. He decided it was better to play a hundred hands and tabulate what percentage of the time he won. Easy enough.

The neurons of most people, even most scientists, wouldn't have made the connection, but in the middle of his century of solitaire, Ulam recognized that he was using the same basic approach as scientists had used in the bomb-building "experiments" in Los Alamos. (The connections are abstract, but the order and layout of the cards were like the random inputs, and the "calculation" was playing the hand.) Discussions soon followed with his calculation-loving friend John von Neumann, another European refugee and Manhattan Project veteran. Ulam and von Neumann realized just how powerful the method might be if they could universalize it and apply it to other situations with multitudes of random variables. In those situations, instead of trying to take into account every complication, every butterfly flapping its wings, they would simply define the problem, pick random inputs, and "plug and chug." Unlike an experiment, the results were not certain. But with enough calculations, they could be pretty darn sure of the probabilities.

In a serendipitous coincidence, Ulam and von Neumann knew the American engineers developing the first electronic computers, such as the ENIAC in Philadelphia. The Manhattan Project "computers" had eventually employed a mechanical punch card system for calculations, but the tireless ENIAC showed more promise for the tedious iterations Ulam and von Neumann envisioned. Historically, the science of probability has its roots in aristocratic casinos, and it's unclear where the nickname for Ulam and von Neumann's approach came from. But Ulam liked to brag that he named it in memory of an uncle who often borrowed money to gamble on the "well-known generator of random integers (between zero and thirty-six) in the Mediterranean principality."

Regardless, Monte Carlo science caught on quickly. It cut down on expensive experiments, and the need for high-quality Monte Carlo simulators drove the early development of computers, pushing them to become faster and more efficient. Symbiotically, the advent of cheap computing meant that Monte Carlo–style experiments, simulations, and models began to take over branches of chemistry, astronomy, and physics, not to mention engineering and stock market analysis. Today, just two generations on, the Monte Carlo method (in various forms) so dominates some fields that many young scientists don't realize how thoroughly they've departed from traditional theoretical or experimental science. Overall, an expedient, a temporary measure—using plutonium and uranium atoms like an abacus to compute nuclear chain reactions—has become an irreplaceable feature of the scientific process. It not only conquered science; it settled down, assimilated, and intermarried with other methods.

In 1949, however, that transformation lay in the future. In those early days, Ulam's Monte Carlo method mostly pushed through the next generation of nuclear weapons. Von Neumann, Ulam, and their ilk would show up at the gymnasium-sized rooms where computers were set up and mysteriously ask if they could run a few programs, starting at 12:00 a.m. and running through the night. The weapons they developed during those dead hours were the "supers," multistage devices a thousand times more powerful than standard A-bombs. Supers used plutonium and uranium to ignite stellar-style fusion in extra-heavy liquid hydrogen, a complicated process that never would have moved beyond secret military reports and into missile silos without digital computation. As historian George Dyson neatly summarized the technological history of that decade, "Computers led to bombs, and bombs led to computers."

After a great struggle to find the proper design for a super, scientists hit upon a dandy in 1952. The obliteration of the Eniwetok atoll in the Pacific Ocean during a test of a super that year showed once again the ruthless brilliance of the Monte Carlo method. Nevertheless, bomb scientists already had something even worse than the supers in the pipeline.

Atomic bombs can get you two ways. A madman who just wants lots of people dead and lots of buildings flattened can stick with a conventional, one-stage fission bomb. It's easier to build, and the big flash-bang should satisfy his need for spectacle, as should aftereffects such as spontaneous tornadoes and the silhouettes of victims seared onto brick walls. But if the madman has patience and wants to do something insidious, if he wants to piss in every well and sow the ground with salt, he'll detonate a cobalt-60 dirty bomb.

Whereas conventional nuclear bombs kill with heat, dirty bombs kill with gamma radiation—malignant X-rays. Gamma rays result from frantic radioactive events, and in addition to burning people frightfully, they dig down into bone marrow and scramble the chromosomes in white blood cells. The cells either die outright, grow cancerous, or grow without constraint and, like humans with gigantism, end up deformed and unable to fight infections. All nuclear bombs release some radiation, but with dirty bombs, radiation is the whole point.

Even endemic leukemia isn't ambitious by some bombs'

standards. Another European refugee who worked on the Manhattan Project, Leo Szilard—the physicist who, to his regret, invented the idea of a self-sustaining nuclear chain reaction around 1933—calculated in 1950 as a wiser, more sober man that sprinkling a tenth of an ounce of cobalt-60 on every square mile of earth would pollute it with enough gamma rays to wipe out the human race, a nuclear version of the cloud that helped kill the dinosaurs. His device consisted of a multistage warhead surrounded by a jacket of cobalt-59. A fission reaction in plutonium would kick off a fusion reaction in hydrogen, and once the reaction started, obviously, the cobalt jacket and everything else would be obliterated. But not before something happened on the atomic level. Down there, the cobalt atoms would absorb neutrons from the fission and fusion, a step called salting. The salting would convert stable cobalt-59 into unsteady cobalt-60, which would then float down like ash.

Lots of other elements emit gamma rays, but there's something special about cobalt. Regular A-bombs can be waited out in underground shelters, since their fallout will vomit up gamma rays immediately and be rendered harmless. Hiroshima and Nagasaki were more or less habitable within days of the 1945 explosions. Other elements absorb extra neutrons like alcoholics do another shot at the bar—they'll get sick someday but not for eons. In that case, after the initial blast, radiation levels would never climb too high.

Cobalt bombs fall devilishly between those extremes, a rare case in which the golden mean is the worst. Cobalt-60 atoms would settle into the ground like tiny land mines. Enough would go off right away to make it necessary to flee, but after five years fully half of the cobalt would still be armed. That steady pulse of gamma shrapnel would mean that cobalt bombs could neither be waited out nor endured. It would take a whole human lifetime for the land to recover. This actually makes cobalt bombs unlikely weapons for war, because the conquering army couldn't occupy the territory. But a scorched-earth madman wouldn't have such qualms.

In his defense, Szilard hoped his cobalt bomb—the first "doomsday device"—would never be built, and no country (as far as the public knows) ever tried. In fact, Szilard conjured up the idea to show the insanity of nuclear war, and people did seize on it. In *Dr. Strangelove*, for example, the Soviet enemies have cobalt bombs. Before Szilard, nuclear weapons were horrifying but not necessarily apocalyptic. After his modest proposal, Szilard hoped that people would know better and give up nukes. Hardly. Soon after the haunting name "promethium" became official, the Soviet Union acquired the bomb, too. The U.S. and Soviet governments soon accepted the less-than-reassuring but aptly named doctrine of MAD, or mutual assured destruction—the idea that, outcomes aside, both sides would lose in any nuclear war. However idiotic as an ethos, MAD did deter people from deploying nukes as tactical weapons. Instead, international tensions hardened into the cold war—a struggle that so infiltrated our society that not even the pristine periodic table escaped its stain.

7

Extending the Table, Expanding the Cold War

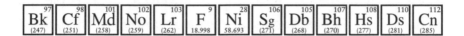

In 1950, a curious notice turned up in the *New Yorker*'s gossipy "Talk of the Town" section:*

> New atoms are turning up with spectacular, if not down-right alarming frequency nowadays, and the University of California at Berkeley, whose scientists have discovered elements 97 and 98, has christened them berkelium and californium respectively.... These names strike us as indicating a surprising lack of public-relations foresight.... California's busy scientists will undoubtedly come up with another atom or two one of these days, and the university...has lost forever the chance of immortalizing itself in the atomic tables with some such sequence as universitium (97), ofium (98), californium (99), berkelium (100).

Not to be outwitted, scientists at Berkeley, led by Glenn Seaborg and Albert Ghiorso, replied that their nomenclature was actually preemptive genius, designed to sidestep "the appalling possibility that after naming 97 and 98 'universitium' and 'ofium,' some New Yorker might follow with the discovery of 99 and 100 and apply the names 'newium' and 'yorkium.'"

The *New Yorker* staff answered, "We are already at work in our office laboratories on 'newium' and 'yorkium.' So far we just have the names."

The letters were fun repartee at a fun time to be a Berkeley scientist. Those scientists were creating the first new elements in our solar system since the supernova kicked everything off billions of years before. Heck, they were outdoing the supernova, making even more elements than the natural ninety-two. No one, least of all them, could foresee how bitter the creation and even the naming of elements would soon become—a new theater for the cold war.

Glenn Seaborg reportedly had the longest *Who's Who* entry ever. Distinguished provost at Berkeley. Nobel Prize–winning chemist. Cofounder of the Pac-10 sports league. Adviser to Presidents Kennedy, Johnson, Nixon, Carter, Reagan, and Bush (George H. W.) on atomic energy and the nuclear arms race. Team leader on the Manhattan Project. Etc., etc. But his first major scientific discovery, which propelled him to those other honors, was the result of dumb luck.

In 1940, Seaborg's colleague and friend, Edwin McMillan, captured a long-standing prize by creating the first transuranic element, which he named neptunium, after the planet beyond uranium's Uranus. Hungry to do more, McMillan further realized that element ninety-three was pretty wobbly and might decay into element ninety-four by spitting off another electron. He searched for evidence of the next element in earnest and kept young Seaborg—a gaunt, twenty-eight-year-old Michigan native who grew up in a Swedish-speaking immigrant colony—apprised of his progress, even discussing techniques while they showered at the gym.

But more was afoot in 1940 than new elements. Once the U.S. government decided to contribute, even clandestinely, to the resistance against the Axis powers in World War II, it began whisking away scientific stars, including McMillan, to work on military projects such as radar. Not prominent enough to be cherry-picked, Seaborg found himself alone in Berkeley with McMillan's equipment and full knowledge of how McMillan had planned to proceed. Hurriedly, fearing it might be their one shot at fame, Seaborg and a colleague amassed a microscopic sample of element ninety-three. After letting the neptunium seep, they sifted through the radioactive sample by dissolving away the excess neptunium, until only a small bit of chemical remained. They proved that the remaining atoms had to be element ninety-four by ripping electron after electron off with a powerful chemical until the atoms held a higher electric charge (+7) than any element ever known. From its very first moments, element ninety-four seemed special. Continuing the march to the edge of the solar system—and under the belief that this was the last possible element that could be synthesized—the scientists named it plutonium.

Suddenly a star himself, Seaborg in 1942 received a summons to go to Chicago and work for a branch of the Manhattan Project. He brought students with him, plus a technician, a sort of super-lackey, named Al Ghiorso. Ghiorso was Seaborg's opposite temperamentally. In pictures, Seaborg invariably appears in a suit, even in the lab, while Ghiorso looks uneasy dressed up, more comfortable in a cardigan and a shirt with the top button undone. Ghiorso wore thick, black-framed glasses and had heavily pomaded hair, and his nose and chin were pointed, a bit like Nixon's. Also unlike Seaborg, Ghiorso chafed at the establishment. (He would have *hated* the Nixon comparison.) A little childishly, he never earned more than a bachelor's degree, not wanting to subject himself to more schooling. Still, prideful, he followed Seaborg to Chicago to escape a monotonous job wiring radioactivity detectors at Berkeley. When he arrived, Seaborg immediately put him to work—wiring detectors.

Nevertheless, the two hit it off. When they returned to Berkeley after the war (both adored the university), they began to produce heavy elements, as the *New Yorker* had it, "with spectacular, if not downright alarming frequency." Other writers have compared chemists who discovered new elements in the 1800s to big-game hunters, thrilling the chemistry-loving masses with every exotic species they bagged. If that flattering description is true, then the stoutest hunters with the biggest elephant guns, the Ernest Hemingway and Theodore Roosevelt of the periodic table, were Ghiorso and Seaborg—who discovered more elements than anyone in history and extended the periodic table by almost one-sixth.

The collaboration started in 1946, when Seaborg, Ghiorso, and others began bombarding delicate plutonium with radioactive particles. This time, instead of using neutron ammunition, they used alpha particles, clusters of two protons and two neutrons. As charged particles, which can be pulled along by dangling a mechanical "rabbit" of the opposite charge in front of their noses, alphas are easier to accelerate to high speeds than mulish neutrons. Plus, when alphas stuck the plutonium, the Berkeley team got two new elements at a stroke, since element ninety-six (plutonium's protons plus two more) decayed into element ninety-five by ejecting a proton.

As discoverers of ninety-five and ninety-six, the Seaborg-Ghiorso team earned the right to name them (an informal tradition soon thrown into angry confusion). They selected americium (pronounced "am-er-EE-see-um"), after America, and curium, after Marie Curie. Departing from his usual stiffness, Seaborg announced the elements not in a scientific journal but on a children's radio show, *Quiz Kids.* A precocious tyke asked Mr. Seaborg if (ha, ha) he'd discovered any new elements lately. Seaborg answered that he had, actually, and encouraged kids listening at home to tell their teachers to throw out the old periodic table. "Judging from the mail I later received from schoolchildren," Seaborg recalled in his autobiography, "their teachers were rather skeptical."

Continuing the alpha-bombing experiments, the Berkeley team discovered berkelium and californium in 1949, as described earlier. Proud of the names, and hoping for a little recognition, they called the Berkeley mayor's office in celebration. The staffers in the mayor's office listened and yawned—neither the mayor nor his staff saw what the big deal about the periodic table was. The city's obtuseness got Ghiorso upset. Before the mayor's snub, he had been advocating for calling element ninety-seven berkelium and making its chemical symbol Bm, because the element had been such a "stinker" to discover. Afterward, he might have been tickled to think that every scatological teenager in the country would see Berkeley represented on the periodic table as "Bm" in school and laugh. (Unfortunately, he was overruled, and the symbol for berkelium became Bk.)

Undeterred by the mayor's cold reception, UC Berkeley kept inking in new boxes on the periodic table, keeping school-chart manufacturers, who had to replace obsolete tables, happy. The team discovered elements ninety-nine and one hundred, einsteinium and fermium, in radioactive coral after a hydrogen bomb test in the Pacific in 1952. But their experimental apex was the creation of element 101.

Because elements grow fragile as they swell with protons, scientists had difficulty creating samples large enough to spray with alpha particles. Getting enough einsteinium (element ninety-nine) to even think about leapfrogging to element 101 required bombarding plutonium for three years. And that was just step one in a veritable Rube Goldberg machine. For each attempt to create 101, the scientists dabbed invisibly tiny bits of einsteinium onto gold foil and pelted it with alpha particles. The irradiated gold trellis then had to be dissolved away, since its radioactivity would interfere with detecting the new element. In previous experiments to find new elements, scientists had poured the sample into test tubes at this point to see what reacted with it, looking for chemical analogues to elements on the periodic table. But with element 101, there weren't enough atoms for that. Therefore, the team had to identify it "posthumously," by looking at what was left over after each atom disintegrated—like piecing a car together from scraps after a bombing.

Such forensic work was doable—except the alpha particle step could be done only in one lab, and the detection could be done only in another, miles away. So for each trial run, while the gold foil was dissolving, Ghiorso waited outside in his Volkswagen, motor running, to courier the sample to the other building. The team did this in the middle of the night, because the sample, if stuck in a traffic jam, might go radioactive in Ghiorso's lap and waste the whole effort. When Ghiorso arrived at the second lab, he dashed up the stairs, and the sample underwent another quick purification before being placed into the latest generation of detectors wired by Ghiorso—detectors he was now proud of, since they were the key apparatus in the most sophisticated heavy-element lab in the world.

The team got the drill down, and one February night in

1955, their work paid off. In anticipation, Ghiorso had wired his radiation detector to the building's fire alarm, and when it finally detected an exploding atom of element 101, the bell shrieked. This happened sixteen more times that night, and with each ring, the assembled team cheered. At dawn, everyone went home drunkenly tired and happy. Ghiorso forgot to unwire his detector, however, which caused some panic among the building's occupants the next morning when a laggard atom of element 101 tripped the alarm one last time.*

Having already honored their home city, state, and country, the Berkeley team suggested mendelevium, after Dmitri Mendeleev, for element 101. Scientifically, this was a no-brainer. Diplomatically, it was daring to honor a Russian scientist during the cold war, and it was not a popular choice (at least domestically; Premier Khrushchev reportedly loved it). But Seaborg, Ghiorso, and others wanted to demonstrate that science rose above petty politics, and at the time, why not? They could afford to be magnanimous. Seaborg would soon depart for Camelot and Kennedy, and under Al Ghiorso's direction, the Berkeley lab chugged along. It practically lapped all other nuclear labs in the world, which were relegated to checking Berkeley's arithmetic. The single time another group, from Sweden, claimed to beat Berkeley to an element, number 102, Berkeley quickly discredited the claim. Instead, Berkeley notched element 102, nobelium (after Alfred Nobel, dynamite inventor and founder of the Nobel Prizes), and element 103, lawrencium (after Berkeley Radiation Laboratory founder and director Ernest Lawrence), in the early 1960s.

Then, in 1964, a second *Sputnik* happened.

Some Russians have a creation myth about their corner of the planet. Way back when, the story goes, God walked the earth, carrying all its minerals in his arms, to make sure they got distributed evenly. This plan worked well for a while. Tantalum went in one land, uranium another, and so on. But when God got to Siberia, his fingers got so cold and stiff, he dropped all the metals. His hands too frostbit to scoop them up, he left them there in disgust. And this, Russians boast, explains their vast stores of minerals.

Despite those geological riches, only two useless elements on the periodic table were discovered in Russia, ruthenium and samarium. Compare that paltry record to the dozens of elements discovered in Sweden and Germany and France. The list of great Russian scientists beyond Mendeleev is similarly barren, at least in comparison to Europe proper. For various reasons—despotic tsars, an agrarian economy, poor schools, harsh weather—Russia just never fostered the scientific genius it might have. It couldn't even get basic technologies right, such as its calendar. Until well past 1900, Russia used a misaligned calendar that Julius Caesar's astrologers had invented, leaving it weeks behind Europe and its modern Gregorian calendar. That lag explains why the "October Revolution" that brought Vladimir Lenin and the Bolsheviks to power in 1917 actually occurred in November.

That revolution succeeded partly because Lenin promised to turn backward Russia around, and the Soviet Politburo insisted that scientists would be first among equals in the new workers' paradise. Those claims held true for a few years, as scientists under Lenin went about their business with little state interference, and some world-class scientists emerged, amply supported by the state. In addition to making scientists happy, money turned out to be powerful propaganda as well. Noting how well-funded even mediocre Soviet colleagues were, scientists outside the Soviet Union hoped (and their hope led them to believe) that at last a powerful government recognized their importance. Even in America, where McCarthyism flourished in the early 1950s, scientists often looked fondly at the Soviet bloc for its material support of scientific progress.

In fact, groups like the ultra-right-wing John Birch Society, founded in 1958, thought the Soviets might even be a little too clever with their science. The society fulminated against the addition of fluoride (ions of fluorine) to tap water to prevent cavities. Aside from iodized salt, fluorinated water is among the cheapest and most effective public health measures ever enacted, enabling most people who drank it to die with their own teeth for the first time in history. To the Birchers, though, fluoridation was linked with sex education and other "filthy Communist plots" to control Americans' minds, a mirrored fun house that led straight from local water officials and health teachers to the Kremlin. Most U.S. scientists watched the anti-science fearmongering of the John Birch Society with horror, and compared to that, the pro-science rhetoric of the Soviet Union must have seemed blissful.

Beneath that skin of progress, though, a tumor had metastasized. Joseph Stalin, who assumed despotic control over the Soviet Union in 1929, had peculiar ideas about science. He divided it—nonsensically, arbitrarily, and poisonously—into "bourgeois" and "proletarian" and punished anyone who practiced the former. For decades, the Soviet agricultural research program was run by a proletarian peasant, the "barefoot scientist" Trofim Lysenko. Stalin practically fell in love with him because Lysenko denounced the regressive idea that living things, including crops, inherit traits and genes from their parents. A proper Marxist, he preached that only the proper social environment mattered (even for plants) and that the Soviet environment would prove superior to the capitalist pig environment. As far as it was possible, he also made biology based on genes "illegal" and arrested or executed dissidents. Somehow Lysenkoism failed to boost crop yields, and the millions of collectivized farmers forced to adopt the doctrine starved. During those famines, an eminent British geneticist gloomily described Lysenko as "completely ignorant of the elementary principles of genetics and plant physiology.... To talk to Lysenko was like trying to explain the differential calculus to a man who did not know his twelve times table."

Moreover, Stalin had no compunction about arresting scientists and forcing them to work for the state in slave labor camps. He shipped many scientists to a notorious nickel works and prison outside Norilsk, in Siberia, where temperatures regularly dropped to –80°F. Though primarily a nickel mine, Norilsk smelled permanently of sulfur, from diesel fumes, and scientists there slaved to extract a good portion of the toxic metals on the periodic table, including arsenic, lead, and cadmium. Pollution was rife, staining the sky, and depending on which heavy metal was in demand, it snowed pink or blue. When all the metals were in demand, it snowed black (and still does sometimes today). Perhaps most creepily, to this day reportedly not one tree grows within thirty miles of the poisonous nickel smelters.* In keeping with the macabre Russian sense of humor, a local joke says that bums in Norilsk, instead of begging for change, collect cups of rain, evaporate the water, and sell the scrap metal for cash. Jokes aside, much of a generation of Soviet science was squandered extracting nickel and other metals for Soviet industry.

An absolute realist, Stalin also distrusted spooky, counterintuitive branches of science such as quantum mechanics and relativity. As late as 1949, he considered liquidating the bourgeois physicists who would not conform to Communist ideology by dropping those theories. He drew back only when a brave adviser pointed out that this might harm the Soviet nuclear weapons program just a bit. Plus, unlike in other areas of science, Stalin's "heart" was never really into purging physicists. Because physics overlaps with weapons research, Stalin's pet, and remains agnostic in response to questions about human nature, Marxism's pet, physicists under Stalin escaped the worst abuses leveled at biologists, psychologists, and economists. "Leave [physicists] in peace," Stalin graciously allowed. "We can always shoot them later."

Still, there's another dimension to the pass that Stalin gave the physical sciences. Stalin demanded loyalty, and the Soviet nuclear weapons program had roots in one loyal subject, nuclear scientist Georgy Flyorov. In the most famous picture of him, Flyorov looks like someone in a vaudeville act: smirking, bald front to crown, a little overweight, with caterpillar eyebrows and an ugly striped tie—like someone who'd wear a squirting carnation in his lapel.

That "Uncky Georgy" look concealed shrewdness. In 1942, Flyorov noticed that despite the great progress German and American scientists had made in uranium fission research in recent years, scientific journals had stopped publishing on the topic. Flyorov deduced that the fission work had become state secrets—which could mean only one thing. In a letter that mirrored Einstein's famous letter to Franklin Roosevelt about starting the Manhattan Project, Flyorov alerted Stalin about his suspicions. Stalin, roused and paranoid, rounded up physicists by the dozens and started them on the Soviet Union's own atomic bomb project. But "Papa Joe" spared Flyorov and never forgot his loyalty.

Nowadays, knowing what a horror Stalin was, it's easy to malign Flyorov, to label him Lysenko, part two. Had Flyorov kept quiet, Stalin might never have known about the nuclear bomb until August 1945. Flyorov's case also evokes another possible explanation for Russia's lack of scientific acumen: a culture of toadyism, which is anathema to science. (During Mendeleev's time, in 1878, a Russian geologist named a mineral that contained samarium, element sixty-two, after his boss, one Colonel Samarski, a forgettable bureaucrat and mining official, and easily the least worthy eponym on the periodic table.)

But Flyorov's case is ambiguous. He had seen many colleagues' lives wasted—including 650 scientists rounded up in one unforgettable purge of the elite Academy of Sciences, many of whom were shot for traitorously "opposing progress." In 1942, Flyorov, age twenty-nine, had deep scientific ambitions and the talent to realize them. Trapped as he was in his homeland, he knew that playing politics was his only hope of advancement. And Flyorov's letter did work. Stalin and his successors were so pleased when the Soviet Union unleashed its own nuclear bomb in 1949 that, eight years later, officials entrusted Comrade Flyorov with his own research lab. It was an isolated facility eighty miles outside Moscow, in the city of Dubna, free from state interference. Aligning himself with Stalin was an understandable, if morally flawed, decision for the young man.

In Dubna, Flyorov smartly focused on "blackboard science"—prestigious but esoteric topics too hard to explain to laypeople and unlikely to ruffle narrow-minded ideologues. And by the 1960s, thanks to the Berkeley lab, finding new elements had shifted from what it had been for centuries—an operation where you got your hands dirty digging through obscure rocks—to a rarefied pursuit in which elements "existed" only as printouts on radiation detectors run by computers (or as fire alarm bells). Even smashing alpha particles into heavy elements was no longer practical, since heavy elements don't sit still long enough to be targets.

Scientists instead reached deeper into the periodic table and tried to fuse lighter elements together. On the surface, these projects were all arithmetic. For element 102, you could theoretically smash magnesium (twelve) into thorium (ninety) or vanadium (twenty-three) into gold (seventy-nine). Few combinations stuck together, however, so scientists had to invest a lot of time in calculations to determine which pairs of elements were worth their money and effort. Flyorov and his colleagues studied hard and copied the techniques of the Berkeley lab. And thanks in large part to him, the Soviet Union had shrugged off its reputation as a backwater in physical science by the late 1950s. Seaborg, Ghiorso, and Berkeley beat the Russians to elements 101, 102, and 103. But in 1964, seven years after the original *Sputnik*, the Dubna team announced it had created element 104 first.

Back in berkelium, californium, anger followed shock. Its pride wounded, the Berkeley team checked the Soviet results and, not surprisingly, dismissed them as premature and sketchy. Meanwhile, Berkeley set out to create element 104 itself—which a Ghiorso team, advised by Seaborg, did in 1969. By that point, however, Dubna had bagged 105, too. Again Berkeley scrambled to catch up, all the while contesting that the Soviets were misreading their own data—a Molotov cocktail of an insult. Both teams produced element 106 in 1974, just months apart, and by that time all the international unity of mendelevium had evaporated.

To cement their claims, both teams began naming "their" elements. The lists are tedious to get into, but it's interesting that the Dubna team, à la berkelium, coined one element dubnium. For its part, Berkeley named element 105 after Otto Hahn and then, at Ghiorso's insistence, named 106 after Glenn

Seaborg—a living person—which wasn't "illegal" but was considered gauche in an irritatingly American way. Across the world, dueling element names began appearing in academic journals, and printers of the periodic table had no idea how to sort through the mess.

Amazingly, this dispute stretched all the way to the 1990s, by which point, to add confusion, a team from West Germany had sprinted past the bickering Americans and Soviets to claim contested elements of their own. Eventually, the body that governs chemistry, the International Union of Pure and Applied Chemistry (IUPAC), had to step in and arbitrate.

IUPAC sent nine scientists to each lab for weeks to sort through innuendos and accusations and to look at primary data. The nine men met for weeks on their own, too, in a tribunal. In the end, they announced that the cold war adversaries would have to hold hands and share credit for each element. That Solomonic solution pleased no one: an element can have only one name, and the box on the table was the real prize.

Finally, in 1995, the nine wise men announced tentatively official names for elements 104 to 109. The compromise pleased Dubna and Darmstadt (home of the West German group), but when the Berkeley team saw seaborgium deleted from the list, they went apoplectic. They called a press conference to basically say, "To hell with you; we're using it in the U.S. of A." A powerful American chemistry body, which publishes prestigious journals that chemists around the world very much like getting published in, backed Berkeley up. This changed the diplomatic situation, and the nine men buckled. When the really final, like-it-or-not list came out in 1996, it included seaborgium at 106, as well as the official names on the table today: rutherfordium (104), dubnium (105), borhium (107), hassium (108), and meitnerium (109). After their win, with a public relations

After decades of bickering with Soviet and West German scientists, a satisfied but frail Glenn Seaborg points toward his namesake element, number 106, seaborgium, the only element ever named for a living person. (Photo courtesy Lawrence Berkeley National Laboratory)

foresight the *New Yorker* had once found lacking, the Berkeley team positioned an age-spotted Seaborg next to a huge periodic table, his gnarled finger pointing only sort of toward seaborgium, and snapped a photo. His sweet smile betrays nothing of the dispute whose first salvo had come thirty-two years earlier and whose bitterness had outlasted even the cold war. Seaborg died three years later.

But a story like this cannot end tidily. By the 1990s, Berkeley chemistry was spent, limping behind its Russian and especially German peers. In remarkably quick succession, between just 1994 and 1996, the Germans stamped out element 110, now named darmstadtium (Ds), after their home base; element 111, roentgenium (Rg), after the great German scientist Wilhelm Röntgen; and element 112, the latest element added to the periodic table, in June 2009, copernicium (Cn).* The German success no doubt explained why Berkeley defended its claims for past glory so tenaciously: it had no prospect of future joy. Nevertheless, refusing to be eclipsed, Berkeley pulled a coup in 1996 by hiring a young Bulgarian named Victor Ninov— who had been instrumental in discovering elements 110 and 112—away from the Germans, to renew the storied Berkeley program. Ninov even lured Al Ghiorso out of semiretirement ("Ninov is as good as a young Al Ghiorso," Ghiorso liked to say), and the Berkeley lab was soon surfing on optimism again.

For their big comeback, in 1999 the Ninov team pursued a controversial experiment proposed by a Polish theoretical physicist who had calculated that smashing krypton (thirty-six) into lead (eighty-two) just might produce element 118. Many denounced the calculation as poppycock, but Ninov, determined to conquer America as he had Germany, pushed for the experiment. Creating elements had grown into a multi-year, multimillion-dollar production by then, not something to undertake on a gamble, but the krypton experiment worked miraculously. "Victor must speak directly to God," scientists joked. Best of all, element 118 decayed immediately, spitting out an alpha particle and becoming element 116, which had never been seen either. With one stroke, Berkeley had scored two elements! Rumors spread on the Berkeley campus that the team would reward old Al Ghiorso with his own element, 118, "ghiorsium."

Except...when the Russians and Germans tried to confirm the results by rerunning the experiments, they couldn't find element 118, just krypton and lead. This null result might have been spite, so part of the Berkeley team reran the experiment themselves. It found nothing, even after months of checking. Puzzled, the Berkeley administration stepped in. When they looked back at the original data files for element 118, they noticed something sickening: there was no data. No proof of element 118 existed until a late round of data analysis, when "hits" suddenly materialized from chaotic 1s and 0s. All signs indicated that Victor Ninov—who had controlled the all-important radiation detectors and the computer software that ran them—had inserted false positives into his data files and passed them off as real. It was an unforeseen danger of the eso-teric approach to extending the periodic table: when elements exist only on computers, one person can fool the world by hijacking the computers.

Mortified, Berkeley retracted the claim for 118. Ninov was fired, and the Berkeley lab suffered major budget cuts, decimating it. To this day, Ninov denies that he faked any data—although, damningly, after his old German lab double-checked his experi-ments there by looking into old data files, it also retracted some (though not all) of Ninov's findings. Perhaps worse, American scientists were reduced to traveling to Dubna to work on heavy elements. And there, in 2006, an international team announced that after smashing ten billion billion calcium atoms into a (gulp) californium target, they had produced three atoms of element 118. Fittingly, the claim for element 118 is contested, but if it holds up—and there's no reason to think it won't—the discov-ery would erase any chance of "ghiorsium" appearing on the periodic table. The Russians are in control, since it happened at their lab, and they're said to be partial to "flyorium."

Part III

PERIODIC CONFUSION: THE EMERGENCE OF COMPLEXITY

8

From Physics to Biology

Glenn Seaborg and Al Ghiorso brought the hunt for unknown elements to a new level of sophistication, but they were hardly the only scientists inking in new spaces on the periodic table. In fact, when *Time* magazine named fifteen U.S. scientists its "Men of the Year" for 1960, it selected as one of the honorees not Seaborg or Ghiorso but the greatest element craftsman of an earlier era, a man who'd nabbed the most slippery and elusive element on the entire table while Seaborg was still in graduate school, Emilio Segrè.

In an attempt to look futuristic, the cover for the issue shows a tiny, throbbing red nucleus. Instead of electrons, it is surrounded by fifteen head shots, all in the same sober, stilted poses familiar to anyone who's ever snickered over the teachers' spread in a yearbook. The lineup included geneticists, astronomers, laser pioneers, and cancer researchers, as well as a mug shot of William Shockley, the jealous transistor scientist and future eugenicist. (Even in this issue, Shockley couldn't help but expound on his theories of race.) Despite the class-picture feel, it was an illustrious crew, and *Time* made the selections to crow about the sudden international dominance of American science. In the first four decades of the Nobel Prize, through

1940, U.S. scientists won fifteen prizes; in the next twenty years, they won forty-two.*

Segrè—who as an immigrant and a Jew also reflected the importance of World War II refugees to America's sudden scientific dominance—was among the older of the fifteen, at fifty-five. His picture appears in the top left quadrant, above and to the left of an even older man—Linus Pauling, age fifty-nine, pictured in the lower middle. The two men helped transform periodic table chemistry and, though not intimate friends, conversed about and exchanged letters on topics of mutual interest. Segrè once wrote Pauling for advice on experiments with radioactive beryllium. Pauling later asked Segrè about the provisional name for element eighty-seven (francium), which Segrè had codiscovered and Pauling wanted to mention in an *Encyclopædia Britannica* article he was writing on the periodic table.

What's more, they could have easily been—in fact, should have been—faculty colleagues. In 1922, Pauling was a hot chemistry recruit out of Oregon, and he wrote a letter to Gilbert Lewis (the chemist who kept losing the Nobel Prize) at the University of California at Berkeley, inquiring about graduate school there. Strangely, Lewis didn't bother answering, so Pauling enrolled at the California Institute of Technology, where he starred as a student and faculty member until 1981. Only later did Berkeley realize it had lost Pauling's letter. Had Lewis seen it, he certainly would have admitted Pauling and then—given Lewis's policy of keeping top graduate students as faculty members—would have bound Pauling to Berkeley for life.

Later, Segrè would have joined Pauling there. In 1938, Segrè became yet another Jewish refugee from fascist Europe when Benito Mussolini bowed to Hitler and sacked all the Jewish professors in Italy. As bad as that was, the circumstances of

Segrè's appointment at Berkeley proved equally humiliating. At the time of his firing, Segrè was on sabbatical at the Berkeley Radiation Lab, a famed cousin of the chemistry department. Suddenly homeless and scared, Segrè begged the director of the "Rad Lab" for a full-time job. The director said yes, of course, but only at a lower salary. He assumed correctly that Segrè had no other options and forced him to accept a 60 percent pay cut, from a handsome $300 per month to $116. Segrè bowed his head and accepted, then sent for his family in Italy, wondering how he would support them.

Segrè got over the slight, and in the next few decades, he and Pauling (especially Pauling) became legends in their respective fields. They remain today two of the greatest scientists most laypeople have never heard of. But a largely forgotten link between them — *Time* certainly didn't bring it up — is that Pauling and Segrè will forever be united in infamy for making two of the biggest mistakes in science history.

Now, mistakes in science don't always lead to baleful results. Vulcanized rubber, Teflon, and penicillin were all mistakes. Camillo Golgi discovered osmium staining, a technique for making the details of neurons visible, after spilling that element onto brain tissue. Even an outright falsehood — the claim of the sixteenth-century scholar and protochemist Paracelsus that mercury, salt, and sulfur were the fundamental atoms of the universe — helped turn alchemists away from a mind-warping quest for gold and usher in real chemical analysis. Serendipitous clumsiness and outright blunders have pushed science ahead all through history.

Pauling's and Segrè's were not those kinds of mistakes. They were hide-your-eyes, don't-tell-the-provost gaffes. In their defense, both men were working on immensely complicated projects that, though grounded in the chemistry of single atoms, vaulted over that chemistry into explaining how systems of atoms should behave. Then again, both men could have avoided their mistakes by studying a little more carefully the very periodic table they helped illuminate.

Speaking of mistakes, no element has been discovered for the "first time" more times than element forty-three. It's the Loch Ness monster of the elemental world.

In 1828, a German chemist announced the discovery of the new elements "polinium" and "pluranium," one of which he presumed was element forty-three. Both turned out to be impure iridium. In 1846, another German discovered "ilmenium," which was actually niobium. The next year someone else discovered "pelopium," which was niobium, too. Element forty-three disciples at last got some good news in 1869, when Mendeleev constructed his periodic table and left a tantalizing gap between forty-two and forty-four. However, though good science itself, Mendeleev's work encouraged a lot of bad science, since it convinced people to look for something they were predisposed to find. Sure enough, eight years later one of Mendeleev's fellow Russians inked "davyium" into box forty-three on the table, even though it weighed 50 percent more than it should have and was later determined to be a mix of three elements. Finally, in 1896 "lucium" was discovered—and discarded as yttrium—just in time for the twentieth century.

The new century proved even crueler. In 1909, Masataka Ogawa discovered "nipponium," which he named for his homeland (*Nippon* in Japanese). All the previous faux forty-threes had been contaminated samples or previously discovered trace elements. Ogawa had actually discovered a new element—just not what he claimed. In his rush to seize element forty-three, he ignored other gaps in the table, and when no one could confirm

his work, he retracted it, ashamed. Only in 2004 did a coun-
tryman reexamine Ogawa's data and determine he had isolated
element seventy-five, rhenium, also undiscovered at the time,
without knowing it. It depends whether you're a half-full or
half-empty kind of person if you think Ogawa would be posthumously pleased to find out he'd discovered at least something,
or even more vexed at his wrenching mistake.

Element seventy-five was discovered unambiguously in 1925
by three more German chemists, Otto Berg and the husband
and wife team of Walter and Ida Noddack. They named it rhenium after the Rhine River. Simultaneously, they announced
yet another stab at element forty-three, which they called
"masurium" after a region of Prussia. Given that nationalism
had destroyed Europe a decade earlier, other scientists did not
look kindly on those Teutonic, even jingoistic names — both
the Rhine and Masuria had been sites of German victories in
World War I. A continent-wide plot rose up to discredit the
Germans. The rhenium data looked solid, so scientists concentrated on the sketchier "masurium" work. According to some
modern scholars, the Germans might have discovered element
forty-three, but the trio's paper contained sloppy mistakes,
such as overestimating by many thousands of times the amount
of "masurium" they had isolated. As a result, scientists already
suspicious of yet another claim for element forty-three declared
the finding invalid.

Only in 1937 did two Italians isolate the element. To do so,
Emilio Segrè and Carlo Perrier took advantage of new work
in nuclear physics. Element forty-three had proved so elusive
until then because virtually every atom of it in the earth's crust
had disintegrated radioactively into molybdenum, element forty-
two, millions of years ago. So instead of sifting through tons
of ore like suckers for a few micro-ounces of it (as Berg and the

Noddacks had), the Italians had an unknowing American colleague make some.

A few years earlier that American, Ernest Lawrence (who once called Berg and the Noddacks' claim for element forty-three "delusional"), had invented an atom smasher called a cyclotron to mass-produce radioactive elements. Lawrence was more interested in creating isotopes of existing elements than in creating new ones, but when Segrè happened to visit Lawrence's lab on a tour of America in 1937, Segrè heard that the cyclotron used replaceable molybdenum parts—at which point his internal Geiger counter went wild. He cagily asked to look at some discarded scraps. Weeks later, at Segrè's request, Lawrence happily flew a few worn-out molybdenum strips to Italy in an envelope. Segrè's hunch proved correct: on the strips, he and Perrier found traces of element forty-three. They had filled the periodic table's most frustrating gap.

Naturally, the German chemists did not abandon their claims for "masurium." Walter Noddack even visited and quarreled with Segrè in the Italian's office—and did so dressed in an intimidating, quasi-military uniform covered with swastikas. This didn't endear him to the short, volatile Segrè, who also faced political pressure on another matter. Officials at the University of Palermo, where Segrè worked, were pushing him to name his new element "panormium," after the Latin for Palermo. Perhaps wary because of the nationalistic debacle over "masurium," Segrè and Perrier chose technetium, Greek for "artificial," instead. It was fitting, if dull, since technetium was the first man-made element. But the name cannot have made Segrè popular, and in 1938 he arranged for a sabbatical abroad at Berkeley, under Lawrence.

There's no evidence Lawrence held a grudge against Segrè for his molybdenum gambit, but it was Lawrence who lowballed

Segrè later that year. In fact, Lawrence blurted out, oblivious to the Italian's feelings, how happy he was to save $184 per month to spend on equipment, like his precious cyclotron. Ouch. This was further proof that Lawrence, for all his skill in securing funds and directing research, was obtuse with people. As often as Lawrence recruited one brilliant scientist, his dictatorial style drove another away. Even a booster of his, Glenn Seaborg, once said that Lawrence's world-renowned and much-envied Rad Lab—and not the Europeans who did—should have discovered artificial radioactivity and nuclear fission, the most momentous discoveries in science at the time. To miss both, Seaborg rued, was "scandalous failure."

Still, Segrè might have sympathized with Lawrence on that last account. Segrè had been a top assistant to the legendary Italian physicist Enrico Fermi in 1934 when Fermi reported to the world (wrongly, it turned out) that by bombarding uranium samples with neutrons, he had "discovered" element ninety-three and other transuranic elements. Fermi long had a reputation as the quickest wit in science, but in this case his snap judgment misled him. In fact, he missed a far more consequential discovery than transuranics: he had actually induced uranium fission years before anyone else and hadn't realized it. When two German scientists contradicted Fermi's results in 1939, Fermi's whole lab was stunned—he had already won a Nobel Prize for this. Segrè felt especially chagrined. His team had been in charge of analyzing and identifying the new elements. Worse, he instantly remembered that he (among others) had read a paper on the possibility of fission in 1934 and had dismissed it as ill conceived and unfounded—a paper by, of all the damned luck, Ida Noddack.*

Segrè—who later became a noted science historian (as well as, incidentally, a noted hunter of wild mushrooms)—wrote about the fission mistake in two books, saying the same terse thing both times: "Fission...escaped us, although it was called specifically to our attention by Ida Noddack, who sent us an article in which she clearly indicated the possibility.... The reason for our blindness is not clear."* (As a historical curiosity, he might also have pointed out that the two people who came closest to discovering fission, Noddack and Irène Joliot-Curie—daughter of Marie Curie—and the person who eventually did discover it, Lise Meitner, were all women.)

Unfortunately, Segrè learned his lesson about the absence of transuranic elements too literally, and he soon had his own solo scandalous failure to account for. Around 1940, scientists assumed that the elements just before and just after uranium were transition metals. According to their arithmetic, element ninety fell in column four, and the first non–naturally occurring element, ninety-three, fell in column seven beneath technetium. But as the modern table shows, the elements near uranium are not transition metals. They sit beneath the rare earths at the bottom of the table and act like rare earths, not like technetium, in chemical reactions. The reason for chemists' blindness back then is clear. Despite their homage to the periodic table, they didn't take periodicity seriously enough. They thought the rare earths were strange exceptions whose quirky, clingy chemistry would never repeat. But it does repeat: uranium and others bury electrons in f-shells just like the rare earths. They must, therefore, jump off the main periodic table at the same point and behave like them in reactions. Simple, at least in retrospect. A year after the bombshell discovery of fission, a colleague down the hall from Segrè decided to try again to find element ninety-three, so he irradiated some uranium in the cyclotron. Believing (for the reasons above) that this new element would act like technetium, he asked Segrè for help, since Segrè had discovered technetium and knew its chemistry better than anyone. Segrè, an eager element hunter, tested the samples. Taking after his quick-witted mentor, Fermi, he announced that they acted like rare earths, not like heavy cousins of technetium. More humdrum nuclear fission, Segrè declared, and he dashed off a paper with the glum title "An Unsuccessful Search for Transuranic Elements."

But while Segrè moved on, the colleague, Edwin McMillan, felt troubled. All elements have unique radioactive signatures, and Segrè's "rare earths" had different signatures than the other rare earths, which didn't make sense. After careful reasoning, McMillan realized that perhaps the samples acted like rare earths because they were chemical cousins of rare earths and diverged from the main periodic table, too. So he and a partner redid the irradiation and chemical tests, cutting Segrè out, and they immediately discovered nature's first forbidden element, neptunium. The irony is too good not to point out. Under Fermi, Segrè had misidentified nuclear fission products as transuranics. "Apparently not learning from that experience," Glenn Seaborg recalled, "once again Segrè saw no need to follow up with careful chemistry." In the exact opposite blunder, Segrè sloppily misidentified transuranic neptunium as a fission product.

Though no doubt furious with himself as a scientist, perhaps as a science historian Segrè could appreciate what happened next. McMillan won the Nobel Prize in Chemistry in 1951 for this work. But the Swedish Academy had rewarded Fermi for discovering the transuranic elements; so rather than admit a mistake, it defiantly rewarded McMillan only for investigating "the *chemistry of* the transuranium elements" (emphasis added). Then again, since careful, mistake-free chemistry had led him to the truth, maybe that wasn't a slight.

*   *   *

If Segrè proved too cocksure for his own good, he was nothing compared to the genius just down I-5 in southern California, Linus Pauling.

After earning his Ph.D. in 1925, Pauling had accepted an eighteen-month fellowship in Germany, then the center of the scientific universe. (Just as all scientists communicate in English today, back then it was de rigueur to speak German.) But what Pauling, still in his twenties, learned about quantum mechanics in Europe soon propelled U.S. chemistry past German chemistry and himself onto the cover of *Time* magazine.

In short, Pauling figured out how quantum mechanics governs the chemical bonds between atoms: bond strength, bond length, bond angle, nearly everything. He was the Leonardo of chemistry—the one who, as Leonardo did in drawing humans, got the anatomical details right for the first time. And since chemistry is basically the study of atoms forming and breaking bonds, Pauling single-handedly modernized the sleepy field. He absolutely deserved one of the great scientific compliments ever paid, when a colleague said Pauling proved "that chemistry could be *understood* rather than being memorized" (emphasis added).

After that triumph, Pauling continued to play with basic chemistry. He soon figured out why snowflakes are six-sided: because of the hexagonal structure of ice. At the same time, Pauling was clearly itching to move beyond straightforward physical chemistry. One of his projects, for instance, determined why sickle-cell anemia kills people: the misshaped hemoglobin in their red blood cells cannot hold on to oxygen. This work on hemoglobin stands out as the first time anyone had traced a disease to a malfunctioning molecule,* and it transformed how doctors thought of medicine. Pauling then, in 1948, while laid up with the flu, decided to revolutionize molecular biology by showing how proteins can form long cylinders called alpha-helixes. Protein function depends largely on protein shape, and Pauling was the first to figure out how the individual bits in proteins "know" what their proper shape is.

In all these cases, Pauling's real interest (besides the obvious benefits to medicine) was in how new properties emerge, almost miraculously, when small, dumb atoms self-assemble into larger structures. The really fascinating angle is that the parts often betray no hint of the whole. Just as you could never guess, unless you'd seen it, that individual carbon, oxygen, and nitrogen atoms could run together into something as useful as an amino acid, you'd have no idea that a few amino acids could fold themselves into all the proteins that run a living being. This work, the study of atomic ecosystems, was a step up in sophistication even from creating new elements. But that jump in sophistication also left more room for misinterpretation and mistakes. In the long run, Pauling's easy success with alpha-helixes proved ironic: had he not blundered with another helical molecule, DNA, he would surely be considered one of the top five scientists ever.

Like most others, Pauling was not interested in DNA until 1952, even though Swiss biologist Friedrich Miescher had discovered DNA in 1869. Miescher did so by pouring alcohol and the stomach juice of pigs onto pus-soaked bandages (which local hospitals gladly gave to him) until only a sticky, goopy, grayish substance remained. Upon testing it, Miescher immediately and self-servingly declared that deoxyribonucleic acid would prove important in biology. Unfortunately, chemical analysis showed high levels of phosphorus in it. Back then, proteins were considered the only interesting part of biochemistry, and since proteins contain zero phosphorus, DNA was judged a vestige, a molecular appendix.*

Only a dramatic experiment with viruses in 1952 reversed that prejudice. Viruses hijack cells by clamping onto them and then, like inverse mosquitoes, injecting rogue genetic information. But no one knew whether DNA or proteins carried that information. So two geneticists used radioactive tracers to tag both the phosphorus in viruses' phosphorus-rich DNA and the sulfur in their sulfur-rich proteins. When the scientists examined a few hijacked cells, they found that radioactive phosphorus had been injected and passed on but the sulfurous proteins had not. Proteins couldn't be the carriers of genetic information. DNA was.*

But what was DNA? Scientists knew a little. It came in long strands, and each strand consisted of a phosphorus-sugar backbone. There were also nucleic acids, which stuck out from the backbone like knobs on a spine. But the shape of the strands and how they linked up were mysteries—important mysteries. As Pauling showed with hemoglobin and alpha-helixes, shape relates intimately to how molecules work. Soon DNA shape became the consuming question of molecular biology.

And Pauling, like many others, assumed he was the only one smart enough to answer it. This wasn't, or at least wasn't only, arrogance: Pauling had simply never been beaten before. So in 1952, with a pencil, a slide rule, and sketchy secondhand data, Pauling sat down at his desk in California to crack DNA. He first decided, incorrectly, that the bulky nucleic acids sat on the outside of each strand. Otherwise, he couldn't see how the molecule fit together. He accordingly rotated the phosphorus-sugar backbone toward the molecule's core. Pauling also reasoned, using the bad data, that DNA was a triple helix. That's because the bad data was taken from desiccated, dead DNA, which coils up differently than wet, live DNA. The strange coiling made the molecule seem more twisted than it is, bound around itself three times. But on paper, this all seemed plausible.

Everything was humming along nicely until Pauling requested that a graduate student check his calculations. The student did and was soon tying himself in knots trying to see where he was wrong and Pauling was right. Eventually, he pointed out to Pauling that it just didn't seem like the phosphate molecules fit, for an elementary reason. Despite the emphasis in chemistry classes on neutral atoms, sophisticated chemists don't think of elements that way. In nature, especially in biology, many elements exist only as ions, charged atoms. Indeed, according to laws Pauling had helped work out, the phosphorus atoms in DNA would always have a negative charge and would therefore repel each other. He couldn't pack three phosphate strands into DNA's core without blowing the damn thing apart.

The graduate student explained this, and Pauling, being Pauling, politely ignored him. It's not clear why Pauling bothered to have someone check him if he wasn't going to listen, but Pauling's reason for ignoring the student is clear. He wanted scientific priority—he wanted every other DNA idea to be considered a knockoff of his. So contra his usual meticulousness, he assumed the anatomical details of the molecule would work themselves out, and he rushed his phosphorus-in, triple-stranded model into print in early 1953.

Meanwhile, across the Atlantic, two gawky graduate students at Cambridge University pored over advance copies of Pauling's paper. Linus Pauling's son, Peter, worked in the same lab as James Watson and Francis Crick* and had provided the paper as a courtesy. The unknown students desperately wanted to solve DNA to make their careers. And what they read in Pauling's paper flabbergasted them: they had built the same model a year before—and had dismissed it, embarrassed, when a colleague had shown what a shoddy piece of work their triple helix was.

During that dressing-down, however, the colleague, Rosalind Franklin, had betrayed a secret. Franklin specialized in X-ray crystallography, which shows the shapes of molecules. Earlier that year, she had examined wet DNA from squid sperm and calculated that DNA was double-stranded. Pauling, while studying in Germany, had studied crystallography, too, and probably would have solved DNA instantly if he'd seen Franklin's good data. (His data for dried-out DNA was also from X-ray crystallography.) However, as an outspoken liberal, Pauling had had his passport revoked by McCarthyites in the U.S. State Department, and he couldn't travel to England in 1952 for an important conference, where he might have heard of Franklin's work. And unlike Franklin, Watson and Crick never shared data with rivals. Instead, they took Franklin's abuse, swallowed their pride, and started working with her ideas. Not long afterward, Watson and Crick saw all their earlier errors reproduced in Pauling's paper.

Shaking off their disbelief, they rushed to their adviser, William Bragg. Bragg had won a Nobel Prize decades before but lately had become bitter about losing out on key discoveries— such as the shape of the alpha-helix—to Pauling, his flamboyant and (as one historian had it) "acerbic and publicity-seeking" rival. Bragg had banned Watson and Crick from working on DNA after their triple-stranded embarrassment. But when they showed him Pauling's boners and admitted they'd continued to work in secret, Bragg saw a chance to beat Pauling yet. He ordered them back to DNA.

First thing, Crick wrote a cagey letter to Pauling asking how that phosphorus core stayed intact—considering Pauling's theories said it was impossible and all. This distracted Pauling with futile calculations. Even while Peter Pauling alerted him that the two students were closing in, Pauling insisted his three-stranded model would prove correct, that he almost had it. Knowing that Pauling was stubborn but not stupid and would see his errors soon, Watson and Crick scrambled for ideas. They never ran experiments themselves, just brilliantly interpreted other people's data. And in 1953, they finally wrested the missing clue from another scientist.

That man told them that the four nucleic acids in DNA (abbreviated A, C, T, and G) always show up in paired proportions. That is, if a DNA sample is 36 percent A, it will always be 36 percent T as well. Always. The same with C and G. From this, Watson and Crick realized that A and T, and C and G, must pair up inside DNA. (Ironically, that scientist had told Pauling the same thing years before on a sea cruise. Pauling, annoyed at his vacation being interrupted by a loudmouth colleague, had blown him off.) What's more, miracle of miracles, those two pairs of nucleic acids fit together snugly, like puzzle pieces. This explained why DNA is packed so tightly together, a tightness that invalidated Pauling's main reason for turning the phosphorus inward. So while Pauling struggled with his model, Watson and Crick turned theirs inside out, so the negative phosphorus ions wouldn't touch. This gave them a sort of twisted ladder—the famed double helix. Everything checked out brilliantly, and before Pauling recovered,* they published this model in the April 25, 1953, issue of *Nature*.

So how did Pauling react to the public humiliation of triple helixes and inverted phosphorus? And to losing out—to his rival Bragg's lab, no less—on the great biological discovery of the century? With incredible dignity. The same dignity all of us should hope we could summon in a similar situation. Pauling admitted his mistakes, conceded defeat, and even promoted Watson and Crick by inviting them to a professional conference he organized in late 1953. Given his stature, Pauling could afford to be magnanimous; his early championing of the double helix proved he was.

The years after 1953 went much better for both Pauling and Segrè. In 1955, Segrè and yet another Berkeley scientist, Owen Chamberlain, discovered the antiproton. Antiprotons are the mirror image of regular protons: they have a negative charge, may travel backward in time, and, scarily, will annihilate any "real" matter, such as you or me, on contact. After the prediction in 1928 that antimatter exists, one type of antimatter, the antielectron (or positron) was quickly and easily discovered in 1932. Yet the antiproton proved to be the elusive technetium of the particle physics world. The fact that Segrè tracked it down after years of false starts and dubious claims is a testament to his persistence. That's why, four years later, his gaffes forgotten, Segrè won the Nobel Prize in Physics.* Fittingly, he borrowed Edwin McMillan's white vest for the ceremony.

After losing out on DNA, Pauling got a consolation prize: an overdue Nobel of his own, in Chemistry in 1954. Typically for him, Pauling then branched out into new fields. Frustrated by his chronic colds, he started experimenting on himself by taking megadoses of vitamins. For whatever reason, the doses seemed to cure him, and he excitedly told others. Eventually, his imprimatur as a Nobel Prize winner gave momentum to the nutritional supplement craze still going strong today, including the scientifically dubious notion (sorry!) that vitamin C can cure a cold. In addition, Pauling—who had refused to work on the Manhattan Project—became the world's leading anti–nuclear weapons activist, marching in protests and penning books with titles such as *No More War!* He even won a second, surprise Nobel Prize in 1962, the Nobel Peace Prize, becoming the only person to win two unshared Nobels. He did, however, share the stage in Stockholm that year with two laureates in medicine or physiology: James Watson and Francis Crick.

9

Poisoner's Corridor: "Ouch-Ouch"

Pauling learned the hardest way that the rules of biology are much more delicate than the rules of chemistry. You can nigh well abuse amino acids chemically and end up with the same bunch of agitated but intact molecules. The fragile and more complex proteins of a living creature will wilt under the same stress, be it heat, acid, or, worst of all, rogue elements. The most delinquent elements can exploit any number of vulnerabilities in living cells, often by masking themselves as life-giving minerals and micronutrients. And the stories of how ingeniously those elements undo life—the exploits of "poisoner's corridor"—provide one of the darker subplots of the periodic table.

The lightest element in poisoner's corridor is cadmium, which traces its notoriety to an ancient mine in central Japan. Miners began digging up precious metals from the Kamioka mines in AD 710. In the following centuries, the mountains of Kamioka yielded gold, lead, silver, and copper as various shoguns and then business magnates vied for the land. But not until twelve hundred years after striking the first lode did miners begin processing cadmium, the metal that made the mines infamous and the cry *"Itai-itai!"* a byword in Japan for suffering.

The Russo-Japanese War of 1904–1905 and World War I

a decade later greatly increased Japanese demand for metals, including zinc, to use in armor, airplanes, and ammunition. Cadmium appears below zinc on the periodic table, and the two metals mix indistinguishably in the earth's crust. To purify the zinc mined in Kamioka, miners probably roasted it like coffee and percolated it with acid, removing the cadmium. Following the environmental regulations of the day, they then dumped the leftover cadmium sludge into streams or onto the ground, where it leeched into the water table.

Today no one would think of dumping cadmium like that. It's become too valuable as a coating for batteries and computer parts, to prevent corrosion. It also has a long history of use in pigments, tanning agents, and solders. In the twentieth century, people even used shiny cadmium plating to line trendy drinking cups. But the main reason no one would dump cadmium today is that it has rather horrifying medical connotations. Manufacturers pulled it from the trendy tankards because hundreds of people fell ill every year when acidic fruit juice, like lemonade, leached the cadmium from the vessel walls. And when rescue workers at Ground Zero after the September 11, 2001, terrorist attacks developed respiratory diseases, some doctors immediately suspected cadmium, among other substances, since the collapse of the World Trade Center towers had vaporized thousands of electronic devices. That assumption was incorrect, but it's telling how reflexively health officials fingered element forty-eight.

Sadly, that conclusion was a reflex because of what happened a century ago near the Kamioka mines. As early as 1912, doctors there noticed local rice farmers being felled by awful new diseases. The farmers came in doubled over with joint and deep bone pain, especially the women, who accounted for forty-nine of every fifty cases. Their kidneys often failed, too, and their bones softened and snapped from the pressure of everyday tasks. One doctor broke a girl's wrist while taking her pulse. The mystery disease exploded in the 1930s and 1940s as militarism overran Japan. Demand for zinc kept the ores and sludge pouring down the mountains, and although the local prefecture (the Japanese equivalent of a state) was removed from actual combat, few areas suffered as much during World War II as those around the Kamioka mines. As the disease crept from village to village, it became known as *itai-itai*, or "ouch-ouch," disease, after the cries of pain that escaped victims.

Only after the war, in 1946, did a local doctor, Noboru Hagino, begin studying itai-itai disease. He first suspected malnutrition as the cause. This theory proved untenable by itself, so he switched his focus to the mines, whose high-tech, Western excavation methods contrasted with the farmers' primitive paddies. With a public health professor's help, Hagino produced an epidemiological map plotting cases of itai-itai. He also made a hydrological map showing where the Jinzu River—which ran through the mines and irrigated the farmers' fields miles away—deposited its runoff. Laid over the top of each other, the two maps look almost identical. After testing local crops, Hagino realized that the rice was a cadmium sponge.

Painstaking work soon revealed the pathology of cadmium. Zinc is an essential mineral, and just as cadmium mixes with zinc in the ground, it interferes with zinc in the body by replacing it. Cadmium also sometimes evicts sulfur and calcium, which explains why it affected people's bones. Unfortunately, cadmium is a clumsy element and can't perform the same biological roles as the others. Even more unfortunately, once cadmium slips into the body, it cannot be flushed out. The malnutrition Hagino suspected at first also played a role. The local diet depended heavily on rice, which lacks essential nutrients, so the farmers' bodies were starved of certain minerals. Cadmium mimicked those minerals well enough that the farmers' cells, in dietary desperation, began to weave it into their organs at even higher rates than they otherwise would have.

Hagino went public with his results in 1961. Predictably and perhaps understandably, the mining company legally responsible, Mitsui Mining and Smelting, denied all wrongdoing (it had only bought the company that had done the damage). To its shame, Mitsui also campaigned to discredit Hagino. When a local medical committee formed to investigate itai-itai, Mitsui made sure that the committee excluded Hagino, the world expert on the disease. Hagino ran an end around by working on newfound cases of itai-itai in Nagasaki, which only bolstered his claims. Eventually, the conscience-stricken local committee, despite being stacked against Hagino, admitted that cadmium might cause the disease. Upon appeal of this wishy-washy ruling, a national government health committee, overwhelmed by Hagino's evidence, ruled that cadmium absolutely causes itai-itai. By 1972, the mining company began paying restitution to 178 survivors, who collectively sought more than 2.3 billion yen annually. Thirteen years later, the horror of element forty-eight still retained such a hold on Japan that when filmmakers needed to kill off Godzilla in the then-latest sequel, *The Return of Godzilla*, the Japanese military in the film deployed cadmium-tipped missiles. Considering that an H-bomb had given Godzilla life, that's a pretty dim view of this element.

Still, itai-itai disease was not an isolated incident in Japan last century. Three other times in the 1900s (twice with mercury, once with sulfur dioxide and nitrogen dioxide), Japanese villagers found themselves victims of mass industrial poisonings. These cases are known as the Big Four Pollution Diseases of Japan. In addition, thousands more suffered from radiation poisoning when the United States dropped a uranium and a plutonium bomb on the island in 1945. But the atomic bombs and three of the Big Four were preceded by the long-silent holocaust near Kamioka. Except it wasn't so silent for the people there. "Itai-itai."

Scarily, cadmium is not even the worst poison among the elements. It sits above mercury, a neurotoxin. And to the right of mercury sit the most horrific mug shots on the periodic table—thallium, lead, and polonium—the nucleus of poisoner's corridor.

This clustering is partly coincidence, but there are legitimate chemical and physical reasons for the high concentration of poisons in the southeast corner. One, paradoxically, is that none of these heavy metals is volatile. Raw sodium or potassium, if ingested, would explode upon contact with every cell inside you, since they react with water. But potassium and sodium are so reactive they never appear in their pure, dangerous form in nature. The poisoner's corridor elements are subtler and can migrate deep inside the body before going off. What's more, these elements (like many heavy metals) can give up different numbers of electrons depending on the circumstances. For example, whereas potassium always reacts as K^+, thallium can be Tl^+ or Tl^{+3}. As a result, thallium can mimic many elements and wriggle into many different biochemical niches.

That's why thallium, element eighty-one, is considered the deadliest element on the table. Animal cells have special ion channels to vacuum up potassium, and thallium rides into the body via those channels, often by skin osmosis. Once inside the body, thallium drops the pretense of being potassium and starts unstitching key amino acid bonds inside proteins and unraveling their elaborate folds, rendering them useless. And unlike cadmium, thallium doesn't stick in the bones or kidneys, but roams like a molecular Mongol horde. Each atom can do an outsized amount of damage.

For these reasons, thallium is known as the poisoner's poison, the element for people who derive an almost aesthetic pleasure from lacing food and drinks with toxins. In the 1960s, a notorious British lad named Graham Frederick Young, after reading sensationalized accounts of serial killers, began experimenting on his family by sprinkling thallium into their teacups and stew pots. He was soon sent to a mental institution but was later, unaccountably, released, at which point he poisoned seventy more people, including a succession of bosses. Only three died, since Young made sure to prolong their suffering with less-than-lethal doses.

Young's victims are hardly alone in history. Thallium has a gruesome record* of killing spies, orphans, and great-aunts with large estates. But rather than relive darker scenes, maybe it's better to recall element eighty-one's single foray into (admittedly morbid) comedy. During its Cuba-obsessed years, the Central Intelligence Agency hatched a plan to powder Fidel Castro's socks with a sort of talcum powder tainted with thallium. The spies were especially tickled that the poison would cause all his hair, including his famous beard, to fall out, which they hoped would emasculate Castro in front of his comrades before killing him. There's no record of why this plan was never attempted.

Another reason thallium, cadmium, and other related elements work so well as poisons is that they stick around for aeons. I don't just mean they accumulate in the body, as cadmium does. Rather, like oxygen, these elements are likely to form stable, near-spherical nuclei that never go radioactive. Therefore, a fair amount of each still survives in the earth's crust. For instance, the heaviest eternally stable element, lead, sits in box eighty-two, a magic number. And the heaviest almost-stable element, bismuth, is its neighbor, in box eighty-three.

Because bismuth plays a surprising role in poisoner's corridor, this oddball element merits a closer look. Some quick bismuth facts: Though a whitish metal with a pinkish hue, bismuth burns with a blue flame and emits yellow fumes. Like cadmium and lead, bismuth has found widespread use in paints and dyes, and it often replaces "red lead" in the crackling fireworks known as dragon's eggs. Also, of the nearly infinite number of possible chemicals you can make by combining elements on the periodic table, bismuth is one of the very few that expands when it freezes. We don't appreciate how bizarre this is because of common ice, which floats on lakes while fish slide around below it. A theoretical lake of bismuth would behave the same way— but almost uniquely so on the periodic table, since solids virtually always pack themselves more tightly than liquids. What's more, that bismuth ice would probably be gorgeous. Bismuth has become a favorite desktop ornament and decorative knick-knack for mineralogists and element nuts because it can form rocks known as hopper crystals, which twist themselves into elaborate rainbow staircases. Newly frozen bismuth might look like Technicolor M. C. Escher drawings come to life.

Bismuth has helped scientists probe the deeper structure of radioactive matter as well. For decades, scientists couldn't resolve conflicting calculations about whether certain elements would last until the end of time. So in 2003, physicists in France took pure bismuth, swaddled it in elaborate shields to block all possible outside interference, and wired detectors around it to try to determine its half-life, the amount of time it would take 50 percent of the sample to disintegrate. Half-life is a common measurement of radioactive elements. If a bucket of one hundred pounds of radioactive element X takes 3.14159 years to drop fifty

The wild, Technicolor swirls of a hopper crystal form when the element bismuth cools into a staircase crystalline pattern. This crystal spans the width of an adult hand. (Ken Keraiff, Krystals Unlimited)

pounds, then the half-life is 3.14159 years. After another 3.14159 years, you'd have twenty-five pounds. Nuclear theory predicted bismuth should have a half-life of twenty billion billion years, much longer than the age of the universe. (You could multiply the age of the universe by itself and get close to the same figure—and still have only a fifty-fifty shot of seeing any given bismuth atom disappear.) The French experiment was more or less a real-life *Waiting for Godot*. But amazingly, it worked. The French scientists collected enough bismuth and summoned enough patience to witness a number of decays. This result proved that instead of being the heaviest stable atom, bismuth will live only long enough to be the final element to go extinct.

(A similarly Beckettesque experiment is running right now

in Japan to determine whether *all* matter will eventually disintegrate. Some scientists calculate that protons, the building blocks of elements, are ever-so-slightly unstable, with a half-life of at least 100 billion trillion trillion years. Undaunted, hundreds of scientists set up a huge underground pool of ultra-pure, ultra-still water deep inside a mineshaft, and they surrounded it with rings of hair-trigger sensors, in case a proton does split on their watch. This is admittedly unlikely, but it's a far more benevolent use of the Kamioka mines than previously.)

It's time to confess the full truth about bismuth, though. It's technically radioactive, yes, and its coordinates on the periodic table imply that element eighty-three should be awful for you. It shares a column with arsenic and antimony, and it crouches among the worst heavy-metal poisons. Yet bismuth is actually benign. It's even medicinal: doctors prescribe it to soothe ulcers, and it's the "bis" in hot-pink Pepto-Bismol. (When people got diarrhea from cadmium-tainted lemonade, the antidote was usually bismuth.) Overall, bismuth is probably the most misplaced element on the table. That statement might cause chagrin among chemists and physicists who want to find mathematical consistency in the table. Really, it's further proof that the table is filled with rich, unpredictable stories if you know where to look.

In fact, instead of labeling bismuth a freakish anomaly, you might consider it a sort of "noble metal." Just as peaceful noble gases cleave the periodic table between two sets of violent—but differently violent—elements, pacific bismuth marks the transition of poisoner's corridor from the conventional retching-and-deep-pain poisons discussed above to the scorching radioactive poisons described below.

Lurking beyond bismuth is polonium, the poisoner's poison of the nuclear age. Like thallium, it makes people's hair fall out, as the world discovered in November 2006 when Alexander

Litvinenko, an ex–KGB agent, was poisoned by polonium in a London sushi restaurant. Past polonium (skipping over, for now, the ultra-rare element astatine) sits radon. As a noble gas, radon is colorless and odorless and reacts with nothing. But as a heavy element, it displaces air, sinks into the lungs, and discharges lethal radioactive particles that lead inevitably to lung cancer—just another way poisoner's corridor can nip you.

Indeed, radioactivity dominates the bottom of the periodic table. It plays the same role the octet rule does for elements near the top: almost everything useful about heavy elements derives from how, and how quickly, they go radioactive. Probably the best way to illustrate this is through the story of a young American who, like Graham Frederick Young, grew obsessed with dangerous elements. But David Hahn wasn't a sociopath. His disastrous adolescence sprang from a desire to help people. He wanted to solve the world's energy crisis and break its addiction to oil so badly—as badly as only a teenager can want something—that this Detroit sixteen-year-old, as part of a clandestine Eagle Scout project gone berserk in the mid-1990s, erected a nuclear reactor in a potting shed in his mother's backyard.*

David started off small, influenced by a book called *The Golden Book of Chemistry Experiments*, written in the same wincingly earnest tone as a 1950s reel-to-reel educational movie. He grew so excited about chemistry that his girlfriend's mother forbade him to speak to guests at her parties because, in the intellectual equivalent of talking with his mouth full, he'd blurt out unappetizing facts about the chemicals in the food they were eating. But his interest wasn't just theoretical. Like many prepubescent chemists, David quickly outgrew his box chemistry set, and he began playing with chemicals violent enough to blow his bedroom walls and carpet to hell. His mother soon banished him to the basement, then the backyard shed, which suited him fine. Unlike many budding scientists, though, David didn't seem to get better at chemistry. Once, before a Boy Scout meeting, he dyed his skin orange when a fake tanning chemical he was working on burped and blew up in his face. And in a move only someone ignorant of chemistry would try, he accidentally exploded a container of purified potassium by tamping it with a screwdriver (a *baaaad* idea). An ophthalmologist was still digging plastic shards out of his eyes months later.

Even after that, the disasters continued, although, in his defense, David did take on increasingly complicated projects, like his reactor. To get started, he applied the little knowledge he'd gleaned about nuclear physics. This knowledge didn't come from school (he was an indifferent, even a poor, student) but from the glowingly pro–nuclear energy pamphlets he wrote away for and from correspondence with government officials who believed sixteen-year-old "Professor Hahn's" ruse about wanting to devise experiments for fictitious students.

Among other things, David learned about the three main nuclear processes—fusion, fission, and radioactive decay. Hydrogen fusion powers stars and is the most powerful and efficient process, but it plays little role in nuclear power on earth, since we can't easily reproduce the temperatures and pressures needed to ignite fusion. David instead relied on uranium fission and the radioactivity of neutrons, which are by-products of fission. Heavier elements such as uranium have trouble keeping positive protons bound in their tiny nuclei, since identical charges repel, so they also pack in neutrons to serve as buffers. When a heavy atom fissions into two lighter atoms of roughly equal size, the lighter atoms require fewer neutron buffers, so they spit the excess neutrons out. Sometimes those neutrons are absorbed by nearby heavy atoms, which become unstable and spit out more neutrons in a chain reaction. In a bomb, you can just let that process happen. Reactors require more touch and control, since you want to string the fission out over a longer period. The main engineering obstacle David faced was that after the uranium atoms fission and release neutrons, the resulting lighter atoms are stable and cannot perpetuate the chain reaction. As a result, conventional reactors slowly die from lack of fuel.

Realizing this—and going obscenely far beyond the atomic energy merit badge he was originally seeking (really)—David decided to build a "breeder reactor," which makes its own fuel through a clever combination of radioactive species. The reactor's initial source of power would be pellets of uranium-233, which readily fissions. (The 233 means the uranium has 141 neutrons plus 92 protons; notice the excess of neutrons.) But the uranium would be surrounded with a jacket of a slightly lighter element, thorium-232. After the fission events, the thorium would absorb a neutron and become thorium-233. Unstable thorium-233 undergoes beta decay by spitting out an electron, and because charges always balance in nature, when it loses a negative electron, thorium also converts a neutron to a positive proton. This addition of a proton shifts it to the next element on the table, protactinium-233. This is also unstable, so the protactinium spits out another electron and transforms into what you started with, uranium-233. Almost magically, you get more fuel just by combining elements that go radioactive in the right way.

David pursued this project on weekends, since he lived only part-time with his mom after his parents' divorce. For safety's sake, he acquired a dentist's lead apron to protect his organs, and anytime he spent a few hours in the backyard shed, he discarded his clothes and shoes. (His mom and stepdad later admitted that they'd noticed him throwing away good clothes and thought it peculiar. They just assumed David was smarter than they were and knew what he was doing.)

Of all the work he did, probably the easiest part of the project was finding the thorium-232. Thorium compounds have extremely high melting points, so they glow extra-bright when heated. They're too dangerous for household lightbulbs, but in industrial settings, especially mines, thorium lamps are common. Instead of wire filaments as wicks, thorium lamps use small mesh nets called mantles, and David ordered hundreds of replacement mantles from a wholesaler, no questions asked. Then, showing improvement in his chemistry, he melted down the mantles into thorium ash with sustained heat from a blowtorch. He treated the ash with $1,000 worth of lithium he had obtained by cutting open batteries with wire cutters. Heating the reactive lithium and ash over a Bunsen burner purified the thorium, giving David a fine jacket for his reactor core.

Unfortunately, or perhaps fortunately, however well David learned radioactive chemistry, the physics escaped him. David first needed uranium-235 to irradiate the thorium and turn it, the thorium, into uranium-233. So he mounted a Geiger counter (a device that registers radioactivity with a *click-click-click-click*) on the dashboard of his Pontiac and cruised rural Michigan, as if he'd just stumble onto a uranium hot spot in the woods. But ordinary uranium is mostly uranium-238, which is a weak source of radioactivity. (Figuring out how to enrich ore by separating uranium-235 and uranium-238, which are chemically identical, was in fact a major achievement of the Manhattan Project.) David eventually scored some uranium ore from a sketchy supplier in the Czech Republic, but again it was ordinary, unenriched uranium, not the volatile kind. Eventually abandoning this approach, Hahn built a "neutron gun" to irradiate his thorium and get the uranium-233 kindling that way, but the gun barely worked.

A few sensational media stories later implied that David

almost succeeded in building a reactor in the shed. In truth, he wasn't close. The legendary nuclear scientist Al Ghiorso once estimated that David started with at least a billion billion times too little fissionable material. David certainly gathered dangerous materials and, depending on his exposure, might well have shortened his life span. But that's easy. There are many ways to poison yourself with radioactivity. There are very, very few ways to harness those elements, with proper timing and controls, to get something useful from them.

Still, the police took no chances when they uncovered David's plan. They found him late one night poking around a parked car and assumed he was a punk stealing tires. After detaining and harassing him, they searched his Pontiac, which he kindly but stupidly warned them was full of radioactive materials. They also found vials of strange powder and hauled him in for questioning. David was savvy enough not to mention the "hot" equipment in the potting shed, most of which he'd already dismantled anyway, scared that he was making too much progress and might leave a crater. While federal agencies wrangled about who was responsible for David—no one had tried to illegally save the world with nuclear power before—the case dragged on for months. In the meantime, David's mother, fearing her house would be condemned, slipped into the laboratory shed one night and hauled almost everything in there to the trash. Months later, officials finally stormed across the neighbors' backyards in hazmat gear to ransack the shed. Even then, the leftover cans and tools showed a thousand times more radioactivity than background levels.

Because he had no malevolent intentions (and September 11 hadn't happened yet), David was mostly let off the hook. He did argue with his parents about his future, however, and after graduating from high school, he enlisted in the Navy, itching to work on nuclear submarines. Given David's history, the

Navy probably had no choice, but instead of letting him work on reactors, it put him on KP and had him swab decks. Unfortunately for him, he never got the chance to work on science in a controlled, supervised setting, where his enthusiasm and nascent talent might, who knows, have done some good.

The denouement of the story of the radioactive Boy Scout is sad. After leaving the military, David drifted back to his suburban hometown and bummed around without much purpose. After a few quiet years, in 2007 police caught him tampering with (actually stealing) smoke detectors from his own apartment building. With David's record, this was a significant offense, since smoke detectors run on a radioactive element, americium. Americium is a reliable source of alpha particles, which can be channeled into an electric current inside detectors. Smoke absorbs the alpha particles, which disrupts the current and sets off the shrill alarm. But David had used americium to make his crude neutron gun, since alpha particles knock neutrons loose from certain elements. Indeed, he'd already been caught once, when he was a Boy Scout, stealing smoke detectors at a summer camp and had been kicked off the grounds.

In 2007, when his mug shot was leaked to the media, David's cherubic face was pockmarked with red sores, as if he had acute acne and had picked every pimple until it bled. But thirty-one-year-old men usually don't come down with acne. The inescapable conclusion was that he'd been reliving his adolescence with more nuclear experiments. Once again, chemistry fooled David Hahn, who never realized that the periodic table is rife with deception. It was an awful reminder that even though the heavy elements along the bottom of the table aren't poisonous in the conventional way, the way that elements in poisoner's corridor are, they're devious enough to ruin a life.

10

Take Two Elements,
Call Me in the Morning

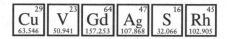

The periodic table is a mercurial thing, and most elements are more complicated than the straightforward rogues of poisoner's corridor. Obscure elements do obscure things inside the body—often bad, but sometimes good. An element toxic in one circumstance can become a lifesaving drug in another, and elements that get metabolized in unexpected ways can provide new diagnostic tools in doctors' clinics. The interplay of elements and drugs can even illuminate how life itself emerges from the unconscious chemical slag of the periodic table.

The reputations of a few elemental medicines extend back a surprisingly long time. Roman officers supposedly enjoyed better health than their grunts because they took their meals on silver platters. And however useless hard currency was in the wild, most pioneer families in early America invested in at least one good silver coin, which spent its Conestoga wagon ride across the wilderness hidden in a milk jug—not for safekeeping, but to keep the milk from spoiling. The noted gentleman astronomer Tycho Brahe, who lost the bridge of his nose in a drunken sword duel in a dimly lit banquet hall in 1564, was even said to have ordered a replacement nose of silver. The metal was fashionable and, more important, curtailed infections. The only drawback was that its obviously metallic color forced Brahe to carry jars of foundation with him, which he was always smoothing over his nasal prosthesis.

Curious archaeologists later dug up Brahe's body and found a green crust on the front of his skull—meaning Brahe had probably worn not a silver but a cheaper, lighter copper nose.* (Or perhaps he switched noses, like earrings, depending on the status of his company.) Either way, copper or silver, the story makes sense. Though both were long dismissed as folk remedies, modern science confirms that those elements have antiseptic powers. Silver is too dear for everyday use, but copper ducts and tubing are standard in the guts of buildings now, as public safety measures. Copper's career in public health began just after America's bicentennial, in 1976, when a plague broke out in a hotel in Philadelphia. Never-before-seen bacteria crept into the moist ducts of the building's air-conditioning system that July, proliferated, and coasted through the vents on a bed of cool air. Within days, hundreds of people at the hotel came down with the "flu," and thirty-four died. The hotel had rented out its convention center that week to a veterans group, the American Legion, and though not every victim belonged, the bug became known as Legionnaires' disease.

The laws pushed through in reaction to the outbreak mandated cleaner air and water systems, and copper has proved the simplest, cheapest way to improve infrastructure. If certain bacteria, fungi, or algae inch across something made of copper, they absorb copper atoms, which disrupt their metabolism (human cells are unaffected). The microbes choke and die after a few hours. This effect—the oligodynamic, or "self-sterilizing," effect—makes metals more sterile than wood or plastic and explains why we have brass doorknobs and metal railings in public places. It also explains why most of the well-handled coins of the U.S. realm contain close to 90 percent copper or (like pennies) are copper-coated.* Copper tubing in air-conditioning ducts cleans out the nasty bugs that fester inside there, too.

Similarly deadly to small wriggling cells, if a bit more quackish, is vanadium, element twenty-three, which also has a curious side effect in males: vanadium is the best spermicide ever devised. Most spermicides dissolve the fatty membrane that surrounds sperm cells, spilling their guts all over. Unfortunately, all cells have fatty membranes, so spermicides often irritate the lining of the vagina and make women susceptible to yeast infections. Not fun. Vanadium eschews any messy dissolving and simply cracks the crankshaft on the sperm's tails. The tails then snap off, leaving the sperm whirling like one-oared rowboats.*

Vanadium hasn't appeared on the market as a spermicide because—and this is a truism throughout medicine—knowing that an element or a drug has desirable effects in test tubes is much different from knowing how to harness those effects and create a safe medicine that humans can consume. For all its potency, vanadium is still a dubious element for the body to metabolize. Among other things, it mysteriously raises and lowers blood glucose levels. That's why, despite its mild toxicity, vanadium water from (as some sites claim) the vanadium-rich springs of Mt. Fuji is sold online as a cure for diabetes.

Other elements have made the transition into effective medicines, like the hitherto useless gadolinium, a potential cancer assassin. Gadolinium's value springs from its abundance of unpaired electrons. Despite the willingness of electrons to bond with other atoms, within their own atoms, they stay maximally far apart. Remember that electrons live in shells, and shells further break down into bunks called orbitals, each of which can accommodate two electrons. Curiously, electrons fill orbitals like patrons find seats on a bus: each electron sits by itself in an orbital until another electron is absolutely forced to double up.* When electrons do condescend to double up, they are picky. They always sit next to somebody with the opposite "spin," a property related to an electron's magnetic field. Linking electrons, spin, and magnets may seem weird, but all spinning charged particles have permanent magnetic fields, like tiny earths. When an electron buddies up with another electron with a contrary spin, their magnetic fields cancel out.

Gadolinium, which sits in the middle of the rare earth row, has the maximum number of electrons sitting by themselves. Having so many unpaired, noncanceling electrons allows gadolinium to be magnetized more strongly than any other element—a nice feature for magnetic resonance imaging (MRI). MRI machines work by slightly magnetizing body tissue with powerful magnets and then flipping the magnets off. When the field releases, the tissue relaxes, reorients itself randomly, and becomes invisible to a magnetic field. Highly magnetic bits like gadolinium take longer to relax, and the MRI machine picks up on that difference. So by affixing gadolinium to tumor-targeting agents—chemicals that seek out and bind only to tumors—doctors can pick tumors out on an MRI scan more easily. Gadolinium basically cranks up the contrast between tumors and normal flesh, and depending on the machine, the tumor will either stand out like a white island in a sea of grayish tissue or appear as an inky cloud in a bright white sky.

Even better, gadolinium might do more than just diagnose tumors. It might also provide doctors with a way to kill those tumors with intense radiation. Gadolinium's array of unpaired electrons allows it to absorb scads of neutrons, which normal body tissue cannot absorb well. Absorbing neutrons turns gadolinium radioactive, and when it goes nuclear, it shreds the tissue around it. Normally, triggering a nano-nuke inside the body is bad, but if doctors can induce tumors to absorb gadolinium, it's sort of an enemy of an enemy thing. As a bonus, gadolinium also inhibits proteins that repair DNA, so the tumor cells cannot rebuild their tattered chromosomes. As anyone who has ever had cancer can attest, a focused gadolinium attack would be a tremendous improvement over chemotherapy and normal cancer radiation, both of which kill cancer cells by scorching everything around them, too. Whereas those techniques are more like firebombs, gadolinium could someday allow oncologists to make surgical strikes without surgery.*

This is not to say that element sixty-four is a wonder drug. Atoms have a way of drifting inside the body, and like any element the body doesn't use regularly, gadolinium has side effects. It causes kidney problems in some patients who cannot flush it out of their systems, and others report that it causes their muscles to stiffen up like early stages of rigor mortis and their skin to harden like a hide, making breathing difficult in some cases. From the looks of it, there's a healthy Internet industry of people claiming that gadolinium (usually taken for an MRI) has ruined their health.

As a matter of fact, the Internet is an interesting place to scout out general claims for obscure medicinal elements. With virtually every element that's not a toxic metal (and even occasionally with those), you can find some alternative medicine site selling it as a supplement.* Probably not coincidentally, you'll also find personal-injury firms on the Internet willing to sue somebody for exposure to nearly every element. So far, the health gurus seem to have spread their message farther and wider than the lawyers, and elemental medicines (e.g., the zinc in lozenges) continue to grow more popular, especially those that have roots as folk remedies. For a century, people gradually replaced folk remedies with prescription drugs, but declining confidence in Western medicine has led some people to self-administer "drugs" such as silver once more.*

Again, there is an ostensible scientific basis for using silver, since it has the same self-sterilizing effects as copper. The difference between silver and copper is that silver, if ingested, colors the skin blue. Permanently. And it's actually worse than that sounds. Calling silvered skin "blue" is easy shorthand. But there's the fun electric blue in people's imaginations when they hear this, and then there's the ghastly gray zombie-Smurf blue people actually turn.

Thankfully, this condition, called argyria, isn't fatal and causes no internal damage. A man in the early 1900s even made a living as "the Blue Man" in a freak show after overdosing on silver nitrate to cure his syphilis. (It didn't work.) In our own times, a survivalist and fierce Libertarian from Montana, the doughty and doughy Stan Jones, ran for the U.S. Senate in 2002 and 2006 despite being startlingly blue. To his credit, Jones had as much fun with himself as the media did. When asked what he told children and adults who pointed at him on the street, he deadpanned, "I just tell them I'm practicing my Halloween costume."

Jones also gladly explained how he contracted argyria. Having his ear to the tin can about conspiracy theories, Jones became obsessed in 1995 with the Y2K computer crash, and especially with the potential lack of antibiotics in the coming apocalypse. His immune system, he decided, had better get ready. So he began to distill a heavy-metal moonshine in his backyard by dipping silver wires attached to 9-volt batteries into tubs of water—a method not even hard-core silver evangelists recommend, since electric currents that strong dissolve far too many silver ions in the bath. Jones drank his stash faithfully for four and a half years, right until Y2K fizzled out in January 2000.

Despite that dud, and despite being gawked at during his serial Senate campaigns, Jones remains unrepentant. He certainly wasn't running for office to wake up the Food and Drug Administration, which in good libertarian fashion intervenes with elemental cures only when they cause acute harm or make promises they cannot possibly keep. A year after losing the 2002 election, Jones told a national magazine, "It's my fault that I overdosed [on silver], but I still believe it's the best antibiotic in the world.... If there were a biological attack on America or if I came down with any type of disease, I'd immediately take it again. Being alive is more important than turning purple."

Stan Jones's advice notwithstanding, the best modern medicines are not isolated elements but complex compounds. Nevertheless, in the history of modern drugs, a few unexpected elements have played an outsized role. This history largely concerns lesser-known heroic scientists such as Gerhard Domagk, but it starts with Louis Pasteur and a peculiar discovery he made about a property of biomolecules called handedness, which gets at the very essence of living matter.

Odds are you're right-handed, but really you're not. You're left-handed. Every amino acid in every protein in your body has a left-handed twist to it. In fact, virtually every protein in every life form that has ever existed is exclusively left-handed. If astrobiologists ever find a microbe on a meteor or moon of Jupiter, almost the first thing they'll test is the handedness of its proteins. If the proteins are left-handed, the microbe is possibly earthly contamination. If they're right-handed, it's certainly alien life.

Pasteur noticed this handedness because he began his career studying modest fragments of life as a chemist. In 1849, at age twenty-six, he was asked by a winery to investigate tartaric acid, a harmless waste product of wine production. Grape seeds and yeast carcasses decompose into tartaric acid and collect as crystals in the dregs of wine kegs. Yeast-born tartaric acid also has a curious property. Dissolve it in water and shine a vertical slit of light through the solution, and the beam will twist clockwise away from the vertical. It's like rotating a dial. Industrial, human-made tartaric acid does nothing like that. A vertical beam emerges true and upright. Pasteur wanted to figure out why.

He determined that it had nothing to do with the chemistry of the two types of tartaric acid. They behaved identically in reactions, and the elemental composition of both was the same. Only when he examined the crystals with a magnifying glass did he notice any difference. The tartaric acid crystals from yeast all twisted in one direction, like tiny, severed left-handed fists. The industrial tartaric acid twisted both ways, a mixture of left- and right-handed fists. Intrigued, Pasteur began the unimaginably tedious job of separating the salt-sized grains into a lefty pile and a righty pile with tweezers. He then dissolved each pile in water and tested more beams of light. Just as he suspected, the yeastlike crystals rotated light clockwise, while the mirror-image crystals rotated light counterclockwise, and exactly the same number of degrees.

Pasteur mentioned these results to his mentor, Jean Baptiste Biot, who had first discovered that some compounds could twist light. The old man demanded that Pasteur show him—then nearly broke down, he was so deeply moved at the elegance of the experiment. In essence, Pasteur had shown that there are two identical but mirror-image types of tartaric acid. More important, Pasteur later expanded this idea to show that life has a strong bias for molecules of only one handedness, or "chirality."*

Pasteur later admitted he'd been a little lucky with this brilliant work. Tartaric acid, unlike most molecules, is easy to see as chiral. In addition, although no one could have anticipated a link between chirality and rotating light, Pasteur had Biot to guide him through the optical rotation experiments. Most serendipitously, the weather cooperated. When preparing the man-made tartaric acid, Pasteur had cooled it on a windowsill. The acid separates into left- and right-handed crystals only below 79°F, and had it been warmer that season, he never would have discovered handedness. Still, Pasteur knew that luck explained just part of his success. As he himself declared, "Chance favors only the prepared mind."

Pasteur was skilled enough for this "luck" to persist throughout his life. Though not the first to do so, he performed an ingenious experiment on meat broth in sterile flasks and proved definitively that air contains no "vitalizing element," no spirit that can summon life from dead matter. Life is built solely, if mysteriously, from the elements on the periodic table. Pasteur also developed pasteurization, a process that heats milk to kill infectious diseases; and, most famously at the time, he saved a young boy's life with his rabies vaccine. For the latter deed, he became a national hero, and he parlayed that fame into the clout he needed to open an eponymous institute outside Paris to further his revolutionary germ theory of disease.

Not quite coincidentally, it was at the Pasteur Institute in the 1930s that a few vengeful, vindictive scientists figured out how the first laboratory-made pharmaceuticals worked—and in doing so hung yet another millstone around the neck of Pasteur's intellectual descendant, the great microbiologist of his era, Gerhard Domagk.

In early December 1935, Domagk's daughter Hildegard tripped down the staircase of the family home in Wuppertal, Germany, while holding a sewing needle. The needle punctured her hand, eyelet first, and snapped off inside her. A doctor extracted the shard, but days later Hildegard was languishing, suffering from a high fever and a brutal streptococcal infection all up and down her arm. As she grew worse, Domagk himself languished and suffered, because death was a frighteningly common outcome for such infections. Once the bacteria began multiplying, no known drug could check their greed.

Except there was one drug—or, rather, one possible drug. It was really a red industrial dye that Domagk had been quietly testing in his lab. On December 20, 1932, he had injected a litter of mice with ten times the lethal dose of streptococcal bacteria. He had done the same with another litter. He'd also injected the second litter with that industrial dye, prontosil, ninety minutes later. On Christmas Eve, Domagk, until that day an insignificant chemist, stole back into his lab to peek. Every mouse in the second litter was alive. Every mouse in the first had died.

That wasn't the only fact confronting Domagk as he kept vigil over Hildegard. Prontosil—a ringed organic molecule that, a little unusually, contains a sulfur atom—had unpredictable properties. Germans at the time believed, a little oddly, that dyes killed germs by turning the germs' vital organs the wrong color. But prontosil, though lethal to microbes inside mice, had no effect on bacteria in test tubes. They swam around happily in the red wash. No one knew why, and because of that ignorance, numerous European doctors had attacked German "chemotherapy," dismissing it as inferior to surgery in treating infection. Even Domagk didn't quite believe in his drug. Between the mouse experiment in 1932 and Hildegard's accident, tentative clinical trials in humans had gone well, but with occasional serious side effects (not to mention that it caused people to flush bright red, like lobsters). Although he was willing to risk the possible deaths of patients in clinical trials for the greater good, risking his daughter was another matter.

In this dilemma, Domagk found himself in the same situation that Pasteur had fifty years before, when a young mother had brought her son, so mangled by a rabid dog he could barely walk, to Pasteur in France. Pasteur treated the boy with a rabies vaccine tested only on animals, and the boy lived.* Pasteur wasn't a licensed doctor, and he administered the vaccine despite the threat of criminal prosecution if it failed. If Domagk failed, he would have the additional burden of having killed a family member. Yet as Hildegard sank further, he likely could not rid his memory of the two cages of mice that Christmas Eve, one teeming with busy rodents, the other still. When Hildegard's doctor announced he would have to amputate her arm, Domagk laid aside his caution. Violating pretty much every research protocol you could draw up, he sneaked some doses of the experimental drug from his lab and began injecting her with the blood-colored serum.

At first Hildegard worsened. Her fever alternately spiked and crashed over the next couple of weeks. Suddenly, exactly three years after her father's mouse experiment, Hildegard stabilized. She would live, with both arms intact.

Though euphoric, Domagk held back mentioning his clandestine experiment to his colleagues, so as not to bias the clinical trials. But his colleagues didn't need to hear about Hildegard to know that Domagk had found a blockbuster—the first genuine antibacterial drug. It's hard to overstate what a revelation this drug was. The world in Domagk's day was modern in many ways. People had quick cross-continental transportation via trains and quick international communication via the telegraph; what they didn't have was much hope of surviving even common infections. With prontosil, plagues that had ravaged human beings since history began seemed conquerable and might even be eradicated. The only remaining question was how prontosil worked.

Not to break my authorial distance, but the following explanation must be chaperoned with an apology. After expounding on the utility of the octet rule, I hate telling you that there are exceptions and that prontosil succeeds as a drug largely because it violates this rule. Specifically, if surrounded by stronger-willed elements, sulfur will farm out all six of its outer-shell electrons and expand its octet into a dozenet. In prontosil's case, the sulfur shares one electron with a benzene ring of carbon atoms, one with a short nitrogen chain, and two each with two greedy oxygen atoms. That's six bonds with twelve electrons, a lot to juggle. And no element but sulfur could pull it off. Sulfur lies in the periodic table's third row, so it's large enough to take on more than eight electrons and bring all those important parts together; yet it's only in the third row and therefore small enough to let everything fit around it in the proper three-dimensional arrangement.

Domagk, primarily a bacteriologist, was ignorant of all that chemistry, and he eventually decided to publish his results so other scientists could help him figure out how prontosil works. But there were tricky business issues to consider. The chemical cartel Domagk worked for, I. G. Farbenindustrie (IGF, the company that later manufactured Fritz Haber's Zyklon B), already sold prontosil as a dye, but it filed for a patent extension on prontosil as a medicine immediately after Christmas in 1932. And with clinical proof that the drug worked well in humans, IGF was fervid about maintaining its intellectual property rights. When Domagk pushed to publish his results, the company forced him to hold back until the medicinal patent on prontosil came through, a delay that earned Domagk and IGF criticism, since people died while lawyers quibbled. Then IGF made Domagk publish in an obscure, German-only periodical, to prevent other firms from finding out about prontosil.

Despite the precaution, and despite prontosil's revolutionary promise, the drug flopped when it hit the market. Foreign doctors continued to harangue about it and many simply didn't believe it could work. Not until the drug saved the life of Franklin Delano Roosevelt Jr., who was struck by a severe strep throat in 1936, and earned a headline in the *New York Times* did prontosil and its lone sulfur atom win any respect. Suddenly, Domagk might as well have been an alchemist for all the money IGF stood to make, and any ignorance about how prontosil worked seemed trifling. Who cared when sales figures jumped fivefold in 1936, then fivefold more the next year.

Meanwhile, scientists at the Pasteur Institute in France had dug up Domagk's obscure journal article. In a froth that was equal parts anti–intellectual property (because they hated how patents hindered basic research) and anti-Teuton (because they hated Germans), the Frenchmen immediately set about busting the IGF patent. (Never underestimate spite as a motivator for genius.)

Prontosil worked as well as advertised on bacteria, but the Pasteur scientists noticed some odd things when they traced its course through the body. First, it wasn't prontosil that fought off bacteria, but a derivative of it, sulfonamide, which mammal cells produce by splitting prontosil in two. This explained instantly why bacteria in test tubes had not been affected: no mammal cells had biologically "activated" the prontosil by cleaving it. Second, sulfonamide, with its central sulfur atom and hexapus of side chains, disrupts the production of folic acid, a nutrient all cells use to replicate DNA and reproduce. Mammals get folic acid from their diets, which means sulfonamide doesn't hobble their cells. But bacteria have to manufacture their own folic acid or they can't undergo mitosis and spread. In effect, then, the Frenchmen proved that Domagk had discovered not a bacteria killer but bacteria birth control!

This breakdown of prontosil was stunning news, and not just medically stunning. The important bit of prontosil, sulfonamide, had been invented years before. It had even been patented in 1909—by I. G. Farbenindustrie*—but had languished because the company had tested it only as a dye. By the mid-1930s, the patent had expired. The Pasteur Institute scientists published their results with undisguised glee, giving everyone in the world a license to circumvent the prontosil patent. Domagk and IGF of course protested that prontosil, not sulfonamide, was the crucial component. But as evidence accumulated against them, they dropped their claims. The company lost millions in product investment, and probably hundreds of millions in profits, as competitors swept in and synthesized other "sulfa drugs."

Despite Domagk's professional frustration, his peers understood what he'd done, and they rewarded Pasteur's heir with the 1939 Nobel Prize in Medicine or Physiology, just seven years after the Christmas mice experiment. But if anything, the Nobel made Domagk's life worse. Hitler hated the Nobel committee for awarding the 1935 Peace Prize to an anti-Nazi journalist and pacifist, and Die Führer had made it basically illegal for any German to win a Nobel Prize. As such, the Gestapo arrested and brutalized Domagk for his "crime." When World War II broke out, Domagk redeemed himself a little by convincing the Nazis (they refused to believe at first) that his drugs could save soldiers suffering from gangrene. But the Allies had sulfa drugs by then, too, and it couldn't have increased Domagk's popularity when his drugs saved Winston Churchill in 1942, a man bent on destroying Germany.

Perhaps even worse, the drug Domagk had trusted to save his daughter's life became a dangerous fad. People demanded sulfonamide for every sore throat and sniffle and soon saw it as some sort of elixir. Their hopes became a nasty joke when quick-buck salesmen in the United States took advantage of this mania by peddling sulfas sweetened with antifreeze. Hundreds died within weeks—further proof that when it comes to panaceas the credulity of human beings is boundless.

Antibiotics were the culmination of Pasteur's discoveries about germs. But not all diseases are germ-based; many have roots in chemical or hormonal troubles. And modern medicine began to address that second class of diseases only after embracing Pasteur's other great insight into biology, chirality. Not long after offering his opinion about chance and the prepared mind, Pasteur said something else that, if not as pithy, stirs a deeper sense of wonder, because it gets at something truly mysterious: what makes life live. After determining that life has a bias toward handedness on a deep level, Pasteur suggested that chirality was the sole "well-marked line of demarcation that at the present can be drawn between the chemistry of dead matter and the chemistry of living matter."* If you've ever wondered what defines life, chemically there's your answer.

Pasteur's statement guided biochemistry for a century, during which doctors made incredible progress in understanding diseases. At the same time, the insight implied that curing diseases, the real prize, would require chiral hormones and chiral biochemicals—and scientists realized that Pasteur's dictum, however perceptive and helpful, subtly highlighted their own ignorance. That is, in pointing out the gulf between the "dead" chemistry that scientists could do in the lab and the living cellular chemistry that supported life, Pasteur simultaneously pointed out there was no easy way to cross it.

That didn't stop people from trying. Some scientists obtained chiral chemicals by distilling essences and hormones from animals, but in the end that proved too arduous. (In the 1920s, two Chicago chemists had to puree several thousand pounds of bull testicles from a stockyard to get a few ounces of the first pure testosterone.) The other possible approach was to ignore Pasteur's distinction and manufacture both right-handed and left-handed versions of biochemicals. This was actually fairly easy to do because, statistically, reactions that produce handed molecules are equally likely to form righties and lefties. The problem with this approach is that mirror-image molecules have different properties inside the body. The zesty odor of lemons and oranges derives from the same basic molecules, one right-handed and one left-handed. Wrong-handed molecules can even destroy left-handed biology. A German drug company in the 1950s began marketing a remedy for morning sickness in pregnant women, but the benign, curative form of the active ingredient was mixed in with the wrong-handed form because the scientists couldn't separate them. The freakish birth defects that followed—especially children born without legs or arms, their hands and feet stitched like turtle flippers to their trunks—made thalidomide the most notorious pharmaceutical of the twentieth century.*

As the thalidomide disaster unfolded, the prospects of

chiral drugs seemed dimmer than ever. But at the same time people were publicly mourning thalidomide babies, a St. Louis chemist named William Knowles began playing around with an unlikely elemental hero, rhodium, in a private research lab at Monsanto, an agricultural company. Knowles quietly circumvented Pasteur and proved that "dead" matter, if you were clever about it, could indeed invigorate living matter.

Knowles had a flat, two-dimensional molecule he wanted to inflate into three dimensions, because the left-handed version of the 3D molecule had shown promising effects on brain diseases such as Parkinson's. The sticking point was getting the proper handedness. Notice that 2D objects cannot be chiral: after all, a flat cardboard cutout of your right hand can be flipped over to make a left hand. Handedness emerges only with the z-axis. But inanimate chemicals in a reaction don't know to make one hand or the other.* They make both, unless they're tricked.

Knowles's trick was a rhodium catalyst. Catalysts speed up chemical reactions to degrees that are hard to comprehend in our poky, everyday human world. Some catalysts improve reaction rates by millions, billions, or even trillions of times. Rhodium works pretty fast, and Knowles found that one rhodium atom could inflate innumerably many of his 2D molecules. So he affixed the rhodium to the center of an already chiral compound, creating a chiral catalyst.

The clever part was that both the chiral catalyst with the rhodium atom and the target 2D molecule were sprawling and bulky. So when they approached each other to react, they did so like two obese animals trying to have sex. That is, the chiral compound could poke its rhodium atom into the 2D molecule only from one position. And from that position, with arms and belly flab in the way, the 2D molecule could unfold into a 3D molecule in only one dimension.

That limited maneuverability during coitus, coupled with rhodium's catalytic ability to fast-forward reactions, meant that Knowles could get away with doing only a bit of the hard work—making a chiral rhodium catalyst—and still reap bushels of correctly handed molecules.

The year was 1968, and modern drug synthesis began at that moment—a moment later honored with a Nobel Prize in Chemistry for Knowles in 2001.

Incidentally, the drug that rhodium churned out for Knowles is levo-dihydroxyphenylalanine, or L-dopa, a compound since made famous in Oliver Sacks's book *Awakenings*. The book documents how L-dopa shook awake eighty patients who'd developed extreme Parkinson's disease after contracting sleeping sickness (*Encephalitis lethargica*) in the 1920s. All eighty were institutionalized, and many had spent four decades in a neurological haze, a few in continuous catatonia. Sacks describes them as "totally lacking energy, impetus, initiative, motive, appetite, affect, or desire...as insubstantial as ghosts, and as passive as zombies...extinct volcanoes."

In 1967, a doctor had had great success in treating Parkinson's patients with L-dopa, a precursor of the brain chemical dopamine. (Like Domagk's prontosil, L-dopa must be biologically activated in the body.) But the right- and left-handed forms of the molecule were knotty to separate, and the drug cost upwards of $5,000 per pound. Miraculously—though without being aware why—Sacks notes that "towards the end of 1968 the cost of L-dopa started a sharp decline." Freed by Knowles's breakthrough, Sacks began treating his catatonic patients in New York not long after, and "in the spring of 1969, in a way... which no one could have imagined or foreseen, these 'extinct volcanoes' erupted into life."

The volcano metaphor is accurate, as the effects of the drug weren't wholly benign. Some people became hyperkinetic, with racing thoughts, and others began to hallucinate or gnaw on things like animals. But these forgotten people almost uniformly preferred the mania of L-dopa to their former listlessness. Sacks recalls that their families and the hospital staff had long considered them "effectively dead," and even some of the victims considered themselves so. Only the left-handed version of Knowles's drug revived them. Once again, Pasteur's dictum about the life-giving properties of proper-handed chemicals proved true.

11

How Elements Deceive

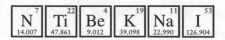

$\mathbf{N}$o one could have guessed that an anonymous gray metal like rhodium could produce anything as wondrous as L-dopa. But even after hundreds of years of chemistry, elements continually surprise us, in ways both benign and not. Elements can muddle up our unconscious, automatic breathing; confound our conscious senses; even, as with iodine, betray our highest human faculties. True, chemists have a good grasp of many features of elements, such as their melting points or abundance in the earth's crust, and the eight-pound, 2,804-page *Handbook of Chemistry and Physics*—the chemists' Koran—lists every physical property of every element to far more decimal places than you'd ever need. On an atomic level, elements behave predictably. Yet when they encounter all the chaos of biology, they continue to baffle us. Even blasé, everyday elements, if encountered in unnatural circumstances, can spring a few mean surprises.

On March 19, 1981, five technicians undid a panel on a simulation spacecraft at NASA's Cape Canaveral headquarters and entered a cramped rear chamber above the engine. A thirty-three-hour "day" had just ended with a perfectly simulated lift-off, and with the space shuttle *Columbia*—the most advanced

space shuttle ever designed—set to launch on its first mission in April, the agency had understandable confidence. The
hard part of their day over, the technicians, satisfied and tired,
crawled into the compartment for a routine systems check. Seconds later, eerily peacefully, they slumped over.

Until that moment, NASA had lost no lives on the ground
or in space since 1967, when three astronauts had burned to
death during training for *Apollo 1*. At the time, NASA, always
concerned about cutting payload, allowed only pure oxygen
to circulate in spacecrafts, not air, which contains 80 percent
nitrogen (i.e., 80 percent deadweight). Unfortunately, as NASA
recognized in a 1966 technical report, "in pure oxygen [flames]
will burn faster and hotter without the dilution of atmospheric
nitrogen to absorb some of the heat or otherwise interfere."
As soon as the atoms in oxygen molecules (O_2) absorb heat,
they dissociate and raise hell by stealing electrons from nearby
atoms, a spree that makes fires burn hotter. Oxygen doesn't
need much provocation either. Some engineers worried that
even static electricity from the Velcro on astronauts' suits
might ignite pure, vigorous oxygen. Nevertheless, the report
concluded that although "inert gas has been considered as a
means of suppressing flammability…inert additives are not
only unnecessary but also increasingly complicated."

Now, that conclusion might be true in space, where atmospheric pressure is nonexistent and just a little interior gas will
keep a spacecraft from collapsing inward. But when training
on the ground, in earth's heavy air, NASA technicians had to
pump the simulators with far more oxygen to keep the walls
from crumpling—which meant far more danger, since even
small fires combust wildly in pure oxygen. When an unex-
plained spark went off one day during training in 1967, fire
engulfed the module and cremated the three astronauts inside.

A disaster has a way of clarifying things, and NASA decided inert gases were necessary, complicated or not, in all shuttles and simulators thereafter. By the 1981 *Columbia* mission, they filled any compartment prone to produce sparks with inert nitrogen (N_2). Electronics and motors work just as well in nitrogen, and if sparks do shoot up, nitrogen—which is locked into molecular form more tightly than oxygen—will smother them. Workers who enter an inert compartment simply have to wear gas masks or wait until the nitrogen is pumped out and breathable air seeps back in—a precaution not taken on March 19. Someone gave the all clear too soon, the technicians crawled into the chamber unaware, and they collapsed as if choreographed. The nitrogen not only prevented their neurons and heart cells from absorbing new oxygen; it pickpocketed the little oxygen cells store up for hard times, accelerating the technicians' demise. Rescue workers dragged all five men out but could revive only three. John Bjornstad was dead, and Forrest Cole died in a coma on April Fools' Day.

In fairness to NASA, in the past few decades nitrogen has asphyxiated miners in caves and people working in underground particle accelerators,* too, and always under the same horror-movie circumstances. The first person to walk in collapses within seconds for no apparent reason. A second and sometimes third person dash in and succumb as well. The scariest part is that no one struggles before dying. Panic never kicks in, despite the lack of oxygen. That might seem incredible if you've ever been trapped underwater. The instinct not to suffocate will buck you to the surface. But our hearts, lungs, and brains actually have no gauge for detecting oxygen. Those organs judge only two things: whether we're inhaling some gas, any gas, and whether we're exhaling carbon dioxide. Carbon dioxide dissolves in blood to form carbonic acid, so as long as we purge CO_2 with each breath and tamp down the acid, our brains will relax. It's an evolutionary kludge, really. It would make more sense to monitor oxygen levels, since that's what we crave. It's easier—and usually good enough—for cells to check that carbonic acid is close to zero, so they do the minimum.

Nitrogen thwarts that system. It's odorless and colorless and causes no acid buildup in our veins. We breathe it in and out easily, so our lungs feel relaxed, and it snags no mental trip wires. It "kills with kindness," strolling through the body's security system with a familiar nod. (It's ironic that the traditional group name for the elements in nitrogen's column, the "pnictogens," comes from a Greek word for "choking" or "strangling.") The NASA workers—the first casualties of the doomed space shuttle *Columbia*, which would disintegrate over Texas twenty-two years later—likely felt light-headed and sluggish in their nitrogen haze. But anyone might feel that way after thirty-three hours of work, and because they could exhale carbon dioxide just fine, little more happened mentally before they blacked out and nitrogen shut down their brains.

Because it has to combat microbes and other living creatures, the body's immune system is more biologically sophisticated than its respiratory system. That doesn't mean it's savvier about avoiding deception. At least, though, with some of the chemical ruses against the immune system, the periodic table deceives the body for its own good.

In 1952, Swedish doctor Per-Ingvar Brånemark was studying how bone marrow produces new blood cells. Having a strong stomach, Brånemark wanted to watch this directly, so he chiseled out holes in the femurs of rabbits and covered the holes with an eggshell-thin titanium "window," which was transparent to strong light. The observation went satisfactorily, and

Brånemark decided to snap off the expensive titanium screens for more experiments. To his annoyance, they wouldn't budge. He gave up on those windows (and the poor rabbits), but when the same thing happened in later experiments—the titanium always locked like a vise onto the femur—he examined the situation a little closer. What he saw made watching juvenile blood cells suddenly seem vastly less interesting and revolutionized the sleepy field of prosthetics.

Since ancient times, doctors had replaced missing limbs with clumsy wooden appendages and peg legs. During and after the industrial revolution, metal prostheses became common, and disfigured soldiers after World War I sometimes even got detachable tin faces—masks that allowed the soldiers to pass through crowds without drawing stares. But no one was able to integrate metal or wood into the body, the ideal solution. The immune system rejected all such attempts, whether made of gold, zinc, magnesium, or chromium-coated pig bladders. As a blood guy, Brånemark knew why. Normally, posses of blood cells surround foreign matter and wrap it in a straitjacket of slick, fibrous collagen. This mechanism—sealing the hunk off and preventing it from leaking—works great with, say, buckshot from a hunting accident. But cells aren't smart enough to distinguish between invasive foreign matter and useful foreign matter, and a few months after implantation, any new appendages would be covered in collagen and start to slip or snap free.

Since this happened even with metals the body metabolizes, such as iron, and since the body doesn't need titanium even in trace amounts, titanium seemed an unlikely candidate for being accepted by the immune system. Yet Brånemark found that for some reason, titanium hypnotizes blood cells: it triggers zero immune response and even cons the body's osteoblasts, its bone-forming cells, into attaching themselves to it as if there was no difference between element twenty-two and actual bone. Titanium can fully integrate itself into the body, deceiving it for its own good. Since 1952, it's been the standard for implanted teeth, screw-on fingers, and replaceable sockets, like the hip socket my mother received in the early 1990s.

Due to cosmically bad luck, arthritis had cleaned out the cartilage in my mother's hip at a young age, leaving bone grinding on bone like a jagged mortar and pestle. She got a full hip replacement at age thirty-five, which meant having a titanium spike with a ball on the end hammered like a railroad tie into her sawed-off femur and the socket screwed into her pelvis. A few months later, she was walking pain-free for the first time in years, and I happily told people she'd had the same surgery as Bo Jackson.

Unfortunately, partly because of her unwillingness to take it easy around her kindergartners, my mother's first hip failed within nine years. The pain and inflammation returned, and another team of surgeons had to cut her open again. It turned out that the plastic component inside the fake hip socket had begun to flake, and her body had dutifully attacked the plastic shards and the tissue around them, covering them with collagen. But the titanium socket anchored to her pelvis hadn't failed and in fact had to be snapped off to fit the new titanium piece. As a memento of her being their youngest two-time hip replacement patient ever, the surgeons at the Mayo Clinic presented my mother with the original socket. She still has it at home, in a manila envelope. It's the size of a tennis ball cut in half, and even today, a decade later, bits of white bone coral are unshakably cemented to the dark gray titanium surface.

Still yet more advanced than our unconscious immune system is our sensory equipment—our touch and taste and smell—the bridges between our physical bodies and our incorporate minds. But it should be obvious by now that new levels of sophistication introduce new and unexpected vulnerabilities into any living system. And it turns out that the heroic deception of titanium is an exception. We trust our senses for true information about the world and for protection from danger, and learning how gullible our senses really are is humbling and a little frightening.

Alarm receptors inside your mouth will tell you to drop a spoonful of soup before it burns your tongue, but, oddly, chili peppers in salsa contain a chemical, capsaicin, that irritates those receptors, too. Peppermint cools your mouth because minty methanol seizes up cold receptors, leaving you shivering as if an arctic blast just blew through. Elements pull similar tricks with smell and taste. If someone spills the tiniest bit of tellurium on himself, he will reek like pungent garlic for weeks, and people will know he's been in a room for hours afterward. Even more baffling, beryllium, element four, tastes like sugar. More than any other nutrient, humans need quick energy from sugar to live, and after millennia of hunting for sustenance in the wild, you'd think we'd have pretty sophisticated equipment to detect sugar. Yet beryllium—a pale, hard-to-melt, insoluble metal with small atoms that look nothing like ringed sugar molecules—lights up taste buds just the same.

This disguise might be merely amusing, except that beryllium, though sweet in minute doses, scales up very quickly to toxic.* By some estimates, up to one-tenth of the human population is hypersusceptible to something called acute beryllium disease, the periodic table equivalent of a peanut allergy. Even for the rest of us, exposure to beryllium powder can scar the lungs with the same chemical pneumonitis that inhaling fine silica causes, as one of the great scientists of all time, Enrico Fermi, found out. When young, the cocksure Fermi used beryllium powder in experiments on radioactive uranium. Beryllium was excellent for those experiments because, when mixed with radioactive matter, it slows emitted particles down. And instead of letting particles escape uselessly into the air, beryllium spikes them back into the uranium lattice to knock more particles loose. In his later years, after moving from Italy to the United States, Fermi grew so bold with these reactions that he started the first-ever nuclear chain reaction, in a University of Chicago squash court. (Thankfully, he was adept enough to stop it, too.) But while Fermi tamed nuclear power, simple beryllium was doing him in. He'd inadvertently inhaled too much of this chemists' confectioner's powder as a young man, and he succumbed to pneumonitis at age fifty-three, tethered to an oxygen tank, his lungs shredded.

Beryllium can lull people who should know better in part because humans have such a screwy sense of taste. Now, some of the five types of taste buds are admittedly reliable. The taste buds for bitter scour food, especially plants, for poisonous nitrogen chemicals, like the cyanide in apple seeds. The taste buds for savory, or umami, lock onto glutamate, the G in MSG. As an amino acid, glutamate helps build proteins, so these taste buds alert you to protein-rich foods. But the taste buds for sweet and sour are easy to fleece. Beryllium tricks them, as does a special protein in the berries of some species of plants. Aptly named miraculin, this protein strips out the unpleasant sourness in foods without altering the overtones of their taste, so that apple cider vinegar tastes like apple cider, or Tabasco sauce like marinara. Miraculin does this both by muting the taste buds for sour and by bonding to the taste buds for sweet and putting them on hair-trigger alert for the stray hydrogen ions (H^+) that acids produce. Along those same lines, people who accidentally inhale hydrochloric or sulfuric acid often recall their teeth aching as if they'd been force-fed raw, extremely sour lemon slices. But as Gilbert Lewis proved, acids are intimately bound up with electrons and other charges. On a molecular level, then, "sour" is simply what we taste when our taste buds open up and hydrogen ions rush in. Our tongues conflate electricity, the flow of charged particles, with sour acids. Alessandro Volta, an Italian count and the inspiration for the eponym "volt," demonstrated this back around 1800 with a clever experiment. Volta had a number of volunteers form a chain and each pinch the tongue of one neighbor. The two end people then put their fingers on battery leads. Instantly, up and down the line, people tasted each other's fingers as sour.

The taste buds for salty also are affected by the flow of charges, but only the charges on certain elements. Sodium triggers the salty reflex on our tongues most strongly, but potassium, sodium's chemical cousin, free rides on top and tastes salty, too. Both elements exist as charged ions in nature, and it's mostly that charge, not the sodium or potassium per se, that the tongue detects. We evolved this taste because potassium and sodium ions help nerve cells send signals and muscles contract, so we'd literally be brain-dead and our hearts would stop without the charge they supply. Our tongues taste other physiologically important ions such as magnesium and calcium* as vaguely salty, too.

Of course, taste being so complicated, saltiness isn't as tidy as that last paragraph implies. We also taste physiologically useless ions that mimic sodium and potassium as salty (e.g., lithium and ammonium). And depending on what sodium and potassium are paired with, even they can taste sweet or sour. Sometimes, as with potassium chloride, the same molecules taste bitter at low concentrations but metamorphose, Wonka-like, into salt licks at high concentrations. Potassium can also shut the tongue down. Chewing raw potassium gymnemate, a chemical in the leaves of the plant *Gymnema sylvestre*, will neuter miraculin, the miracle protein that turns sour into sweet. In fact, after chewing potassium gymnemate, the cocaine-like rush the tongue and heart usually get from glucose or sucrose or fructose reportedly fizzles out: piles of raw sugar heaped on the tongue taste like so much sand.*

All of this suggests that taste is a frighteningly bad guide to surveying the elements. Why common potassium deceives us is strange, but perhaps being overeager and over-rewarding our brain's pleasure centers are good strategies for nutrients. As for beryllium, it deceives us probably because no human being ever encountered pure beryllium until a chemist isolated it in Paris after the French Revolution, so we didn't have time to evolve a healthy distaste for it. The point is that, at least partially, we're products of our environment, and however good our brains are at parsing chemical information in a lab or designing chemistry experiments, our senses will draw their own conclusions and find garlic in tellurium and powdered sugar in beryllium.

Taste remains one of our primal pleasures, and we should marvel at its complexity. The primary component of taste, smell, is the only sense that bypasses our logical neural processing and connects directly to the brain's emotional centers. And as a combination of senses, touch and smell, taste digs deeper into our emotional reservoirs than our other senses do alone. We kiss with our tongues for a reason. It's just that when it comes to the periodic table, it's best to keep our mouths shut.

A live body is so complicated, so butterfly-flaps-its-wings-in-Brazil chaotic, that if you inject a random element into your bloodstream or liver or pancreas, there's almost no telling what will happen. Not even the mind or brain is immune. The highest faculties of human beings—our logic, wisdom, and judgment—are just as vulnerable to deception with elements such as iodine.

Perhaps this shouldn't be a surprise, since iodine has deception built into its chemical structure. Elements tend to get increasingly heavy across rows from left to right, and Dmitri Mendeleev decreed in the 1860s that increasing atomic weight drives the table's periodicity, making increasing atomic weight a universal law of matter. The problem is that universal laws of nature cannot have exceptions, and Mendeleev's craw knew of a particularly intractable exception in the bottom right-hand corner of the table. For tellurium and iodine to line up beneath similar elements, tellurium, element fifty-two, must fall to the left of iodine, element fifty-three. But tellurium outweighs iodine, and it kept stubbornly outweighing it no matter how many times Mendeleev fumed at chemists that their weighing equipment must be deceiving them. Facts is facts.

Nowadays this reversal seems a harmless chemical ruse, a humbling joke on Mendeleev. Scientists know of four pair reversals among the ninety-two natural elements today— argon-potassium, cobalt-nickel, iodine-tellurium, and thorium-protactinium—as well as a few among the ultraheavy, man-made elements. But a century after Mendeleev, iodine got caught up in a larger, more insidious deception, like a three-card monte hustler mixed up in a Mafia hit. You see, a rumor persists to this day among the billion people in India that Mahatma Gandhi, that sage of peace, absolutely hated iodine. Gandhi probably detested uranium and plutonium, too, for the bombs they enabled, but according to modern disciples of Gandhi who want to appropriate his powerful legend, he reserved a special locus of hatred in his heart for element fifty-three.

In 1930, Gandhi led the Indian people in the famous Salt

March to Dandi, to protest the oppressive British salt tax. Salt was one of the few commodities an endemically poor country such as India could produce on its own. People just gathered seawater, let it evaporate, and sold the dry salt on the street from burlap sacks. The British government's taxing of salt production at 8.2 percent was tantamount in greed and ridiculousness to charging bedouins for scooping sand or Eskimos for making ice. To protest this, Gandhi and seventy-eight followers left for a 240-mile march on March 12. They picked up more and more people at each village, and by the time the swelling ranks arrived in the coastal town of Dandi on April 6, they formed a train two miles long. Gandhi gathered the throng around him for a rally, and at its climax he scooped up a handful of saline-rich mud and cried, "With this salt I am shaking the foundation of the [British] Empire!" It was the subcontinent's Boston Tea Party. Gandhi encouraged everyone to make illegal, untaxed salt, and by the time India gained independence seventeen years later, so-called common salt was indeed common in India.

The only problem was that common salt contains little iodine, an ingredient crucial to health. By the early 1900s, Western countries had figured out that adding iodine to the diet is the cheapest and most effective health measure a government can take to prevent birth defects and mental retardation. Starting with Switzerland in 1922, many countries made iodized salt mandatory, since salt is a cheap, easy way to deliver the element, and Indian doctors realized that, with India's iodine-depleted soil and catastrophically high birthrate, they could save millions of children from crippling deformities by iodizing their salt, too.

But even decades after Gandhi's march to Dandi, salt production was an industry by the people, for the people, and iodized salt, which the West pushed on India, retained a whiff of colonialism. As the health benefits became clearer and India modernized, bans on non-iodized salt did spread among Indian state governments between the 1950s and 1990s, but not without dissent. In 1998, when the Indian federal government forced three holdout states to ban common salt, there was a backlash. Mom-and-pop salt makers protested the added processing costs. Hindu nationalists and Gandhians fulminated against encroaching Western science. Some hypochondriacs even worried, without any foundation, that iodized salt would spread cancer, diabetes, tuberculosis, and, weirdly, "peevishness." These opponents worked frantically, and just two years later—with the United Nations and every doctor in India gaping in horror—the prime minister repealed the federal ban on common salt. This technically made common salt legal in only three states, but the move was interpreted as de facto approval. Iodized salt consumption plummeted 13 percent nationwide. Birth defects climbed in tandem.

Luckily, the repeal lasted only until 2005, when a new prime minister again banned common salt. But this hardly solves India's iodine problem. Resentment in Gandhi's name still makes people seethe. The United Nations, hoping to inculcate a love of iodine in a generation with less of a tie to Gandhi, has encouraged children to smuggle salt from their home kitchens to school. There, they and their teachers play chemistry lab by testing for iodine deficiencies. But it's been a losing battle. Although it would cost India just a penny per person per year to produce enough iodized salt for its citizens, the costs of transporting salt are high, and half the country—half a billion people—cannot currently get iodized salt regularly. The consequences are grim, even beyond birth defects. A lack of trace iodine causes goiter, an ugly swelling of the thyroid gland in the neck. If the deficiency persists, the thyroid gland shrivels up. Since the thyroid regulates the production and release of hormones, including brain hormones, the body cannot run smoothly without it. People can quickly lose mental faculties and even regress to mental retardation.

English philosopher Bertrand Russell, another prominent twentieth-century pacifist, once used those medicinal facts about iodine to build a case against the existence of immortal souls. "The energy used in thinking seems to have a chemical origin...," he wrote. "For instance, a deficiency of iodine will turn a clever man into an idiot. Mental phenomena seem to be bound up with material structure." In other words, iodine made Russell realize that reason and emotions and memories depend on material conditions in the brain. He saw no way to separate the "soul" from the body, and concluded that the rich mental life of human beings, the source of all their glory and much of their woe, is chemistry through and through. We're the periodic table all the way down.

Part IV

THE ELEMENTS OF
HUMAN CHARACTER

12

Political Elements

The human mind and brain are the most complex structures known to exist. They burden humans with strong, complicated, and often contradictory desires, and even something as austere and scientifically pure as the periodic table reflects those desires. Fallible human beings constructed the periodic table, after all. Even more than that, the table is where the conceptual meets the grubby, where our aspirations to know the universe—humankind's noblest faculties—have to interact with the material matter that makes up our world—the stuff of our vices and limitations. The periodic table embodies our frustrations and failures in every human field: economics, psychology, the arts, and—as the legacy of Gandhi and the trials of iodine prove—politics. No less than a scientific, there's a social history of the elements.

That history can best be traced through Europe, starting in a country that was as much of a pawn for colonial powers as Gandhi's India. Like a cheap theater set, Poland has been called a "country on wheels" for all its exits and entrances on the world stage. The empires surrounding Poland—Russia, Austria, Hungary, Prussia, Germany—have long held war scrimmages on this flat, undefended turf and have taken turns carving up "God's playground" politically. If you randomly select a map from any year in the past five centuries, the odds are good that Polska (Poland) will be missing.

Fittingly, Poland did not exist when one of the most illustrious Poles ever, Marie Skłodowska, was born in Warsaw in 1867, just as Mendeleev was constructing his great tables. Russia had swallowed Warsaw up four years earlier after a doomed (as most Polish ones were) revolt for independence. Tsarist Russia had backward views on educating women, so the girl's father tutored her himself. She showed aptitude in science as an adolescent, but also joined bristly political groups and agitated for independence. After demonstrating too often against the wrong people, Skłodowska found it prudent to move to Poland's other great cultural center, Krakow (which at the time, *sigh*, was Austrian). Even there, she could not obtain the science training she coveted. She finally moved to the Sorbonne in faraway Paris. She planned to return to her homeland after she earned a Ph.D., but upon falling in love with Pierre Curie, she stayed in France.

In the 1890s, Marie and Pierre Curie began perhaps the most fruitful collaboration in science history. Radioactivity was the brilliant new field of the day, and Marie's work on uranium, the heaviest natural element, provided a crucial early insight: its chemistry was separate from its physics. Atom for atom, pure uranium emitted just as many radioactive rays as uranium in minerals because the electron bonds between a uranium atom and the atoms surrounding it (its chemistry) did not affect if or when its nucleus went radioactive (its physics). Scientists no longer had to examine millions of chemicals and tediously measure the radioactivity of each (as they must do to figure out melting points, for instance). They needed to study only the ninety-some elements on the periodic table. This vastly simplified the field, clearing away the distracting cobwebs and revealing the wooden beam holding up the edifice. The Curies shared the 1903 Nobel Prize in Physics for making this discovery.

During this time, life in Paris satisfied Marie, and she had a daughter, Irène, in 1897. But she never stopped viewing herself as Polish. Indeed, Curie was an early example of a species whose population exploded during the twentieth century—the refugee scientist. Like any human activity, science has always been filled with politics—with backbiting, jealousy, and petty gambits. Any look at the politics of science wouldn't be complete without examples of those. But the twentieth century provides the best (i.e., the most appalling) historical examples of how the sweep of empires can also warp science. Politics marred the careers of probably the two greatest women scientists ever, and even purely scientific efforts to rework the periodic table opened rifts between chemists and physicists. More than anything, politics proved the folly of scientists burying their heads in lab work and hoping the world around them figured out its problems as tidily as they did their equations.

Not long after her Nobel Prize, Curie made another fundamental discovery. After running experiments to purify uranium, she noticed, curiously, that the leftover "waste" she normally discarded was three hundred times more radioactive than uranium. Hopeful that the waste contained an unknown element, she and her husband rented a shed once used to dissect corpses and began boiling down thousands of pounds of pitchblende, a uranium ore, in a cauldron and stirring it with "an iron rod almost as big as myself," she reported, just to get enough grams of the residue to study properly. It took years of oppressively tedious work, but the labor culminated in *two* new elements—and was consummated with, since they were elements far, far more radioactive than anything known before, another Nobel Prize in 1911, this one in chemistry.

It may seem odd that the same basic work was recognized in different prize categories, but back then the distinction between fields in atomic science wasn't as clear as it is today. Many early winners in both chemistry and physics won for work related to the periodic table, since scientists were still sorting the table out. (Only by the time Glenn Seaborg and his crew created element ninety-six and named it curium in Marie's honor was the work considered firmly chemistry.) Nevertheless, no one but Marie Curie emerged from that early era with more than one Nobel.

As discoverers of the new elements, the Curies earned the right to name them. To capitalize on the sensation these strange radioactive metals caused (not least because one of the discoverers was a woman), Marie called the first element they isolated polonium—from the Latin for Poland, Polonia—after her nonexistent homeland. No element had been named for a political cause before, and Marie assumed that her daring choice would command worldwide attention and invigorate the Polish struggle for independence. Not quite. The public blinked and yawned, then gorged itself on the salacious details of Marie's personal life instead.

First, tragically, a street carriage ran over and killed Pierre* in 1906 (which is why he didn't share the second Nobel Prize; only living people are eligible for the prize). A few years later, in a country still seething over the Dreyfus Affair (when the French army fabricated evidence of spying against a Jewish officer named Dreyfus and convicted him of treason), the prestigious French Academy of Sciences rejected Marie for admission for being a woman (which was true) and a suspected Jew (which wasn't). Soon after, she and Paul Langevin, her scientific colleague—and, it turned out, lover—attended a conference in Brussels together. Miffed at their holiday, Mrs. Langevin sent Paul and Marie's love letters to a scurrilous newspaper, which published all the juicy bits. A humiliated Langevin ended up fighting pistol duels to salvage Curie's honor, though no one was shot. The only casualty resulted when Mrs. Langevin KO'd Paul with a chair.

The Langevin scandal broke in 1911, and the Swedish Academy of Sciences debated nixing Curie's nomination for her second Nobel Prize, fearing the political fallout of attaching itself to her. It decided it couldn't in good scientific conscience do that, but it did ask her not to attend the ceremony in her honor. She flauntingly showed up anyway. (Marie had a habit of flouting convention. Once, while visiting an eminent male scientist's home, she ushered him and a second man into a dark closet to show off a vial of a radioactive metal that glowed in the dark. Just as their eyes adjusted, a curt knock interrupted them. One of the men's wives was aware of Curie's femme fatale reputation and thought they were taking a little long in there.)

Marie found a slight reprieve from her rocky personal life* when the cataclysm of World War I and the breakup of European empires resurrected Poland, which enjoyed its first taste of independence in centuries. But naming her first element after Poland contributed nothing to the effort. In fact, it turned out to have been a rash decision. As a metal, polonium is useless. It decays so quickly it might have been a mocking metaphor for Poland itself. And with the demise of Latin, its name calls to mind not Polonia but Polonius, the doddering fool from *Hamlet*. Worse, the second element, radium, glows a translucent green and soon appeared in consumer products worldwide. People even drank radium-infused water from radium-lined crocks called Revigators as a health tonic. (A competing company, Radithor, sold individual, pre-seeped bottles of radium and thorium water.)* In all, radium overshadowed its brother and caused exactly the sensation Curie had hoped for with

The trendy Revigator, a pottery crock lined with nuclear radium. Users filled the flask with water, which turned radioactive after a night's soak. Instructions suggested drinking six or more refreshing glasses a day. (National Museum of Nuclear Science and History)

polonium. Moreover, polonium has been linked to lung cancer from cigarettes, since tobacco plants absorb polonium excessively well and concentrate it in their leaves. Once incinerated and inhaled, the smoke ravishes lung tissue with radioactivity. Of all the countries in the world, only Russia, the many-time conqueror of Poland, bothers to manufacture polonium anymore. That's why when ex–KGB spy Alexander Litvinenko ate polonium-laced sushi and appeared in videos looking like a teenage leukemia victim, having lost all his hair, even his eyebrows, his former Kremlin employers became the prime suspects.

Historically, only a single case of acute polonium poisoning has approached the drama of Litvinenko's—that of Irène

Joliot-Curie, Marie's slender, sad-eyed daughter. A brilliant scientist herself, Irène and her husband, Frédéric Joliot-Curie, picked up on Marie's work and soon one-upped her. Rather than just finding radioactive elements, Irène figured out a method for converting tame elements into artificially radioactive atoms by bombarding them with subatomic particles. This work led to her own Nobel Prize in 1935. Unfortunately, Joliot-Curie relied on polonium as her atomic bombardier. And one day in 1946, not long after Poland had been wrested from Nazi Germany, only to be taken over as a puppet of the Soviet Union, a capsule of polonium exploded in her laboratory, and she inhaled Marie's beloved element. Though spared Litvinenko's public humiliation, Joliot-Curie died of leukemia in 1956, just as her mother had twenty-two years before.

The helpless death of Irène Joliot-Curie proved doubly ironic because the cheap, artificial radioactive substances she made possible have since become crucial medical tools. When swallowed in small amounts, radioactive "tracers" light up organs and soft tissue as effectively as X-rays do bones. Virtually every hospital in the world uses tracers, and a whole branch of medicine, radiology, deals exclusively in that line. It's startling to learn, then, that tracers began as no more than a stunt by a graduate student—a friend of Joliot-Curie's who sought revenge on his landlady.

In 1910, just before Marie Curie collected her second Nobel Prize for radioactivity, young György Hevesy arrived in England to study radioactivity himself. His university's lab director in Manchester, Ernest Rutherford, immediately assigned Hevesy the Herculean task of separating out radioactive atoms from nonradioactive atoms inside blocks of lead. Actually, it turned out to be not Herculean but impossible. Rutherford had assumed the radioactive atoms, known as radium-D, were a unique substance. In fact, radium-D was radioactive lead and therefore could not be separated chemically. Ignorant of this, Hevesy wasted two years tediously trying to tease lead and radium-D apart before giving up.

Hevesy—a bald, droopy-cheeked, mustached aristocrat from Hungary—also faced domestic frustrations. Hevesy was far from home and used to savory Hungarian food, not the English cooking at his boardinghouse. After noticing patterns in the meals served there, Hevesy grew suspicious that, like a high school cafeteria recycling Monday's hamburgers into Thursday's beef chili, his landlady's "fresh" daily meat was anything but. When confronted, she denied this, so Hevesy decided to seek proof.

Miraculously, he'd achieved a breakthrough in the lab around that time. He still couldn't separate radium-D, but he realized he could flip that to his advantage. He'd begun musing over the possibility of injecting minute quantities of dissolved lead into a living creature and then tracing the element's path, since the creature would metabolize the radioactive and non-radioactive lead the same way, and the radium-D would emit beacons of radioactivity as it moved. If this worked, he could actually track molecules inside veins and organs, an unprecedented degree of resolution.

Before he tried this on a living being, Hevesy decided to test his idea on the tissue of a nonliving being, a test with an ulterior motive. He took too much meat at dinner one night and, when the landlady's back was turned, sprinkled "hot" lead over it. She gathered his leftovers as normal, and the next day Hevesy brought home a newfangled radiation detector from his lab buddy, Hans Geiger. Sure enough, when he waved it over that night's goulash, Geiger's counter went furious: *click-*

click-click-click. Hevesy confronted his landlady with the evidence. But, being a scientific romantic, Hevesy no doubt laid it on thick as he explained the mysteries of radioactivity. In fact, the landlady was so charmed to be caught so cleverly, with the latest tools of forensic science, that she didn't even get mad. There's no historical record of whether she altered her menu, however.

Soon after discovering elemental tracers, Hevesy's career blossomed, and he continued to work on projects that straddled chemistry and physics. Yet those two fields were clearly diverging, and most scientists picked sides. Chemists remained interested in whole atoms bonding to one another. Physicists were fascinated with the individual parts of atoms and with a new field called quantum mechanics, a bizarre but beautiful way to talk about matter. Hevesy left England in 1920 to study in Copenhagen with Niels Bohr, a major quantum physicist. And it was in Copenhagen that Bohr and Hevesy unwittingly opened the crack between chemistry and physics into a real political rift.

In 1922, the box for element seventy-two on the periodic table stood blank. Chemists had figured out that the elements between fifty-seven (lanthanum) and seventy-one (lutetium) all had rare earth DNA. Element seventy-two was ambiguous. No one knew whether to tack it onto the end of the hard-to-separate rare earths—in which case element hunters should sift through samples of the recently discovered lutetium—or to provisionally classify it as a transition metal, deserving its own column.

According to lore, Niels Bohr, alone in his office, constructed a nearly euclidean proof that element seventy-two was *not* a lutetium-like rare earth. Remember that the role of electrons in chemistry was not well-known, and Bohr supposedly based his proof on the strange mathematics of quantum mechanics, which says that elements can hide only so many electrons in their inner shells. Lutetium and its f-shells had electrons stuffed into every sleeve and cranny, and Bohr reasoned that the next element had no choice but to start putting electrons on display and act like a proper transition metal. Therefore, Bohr dispatched Hevesy and physicist Dirk Coster to scrutinize samples of zirconium—the element above number seventy-two on the table and its probable chemical analogue. In perhaps the least-sweat discovery in periodic table history, Hevesy and Coster found element seventy-two on their first attempt. They named it hafnium, from Hafnia, the Latin name for Copenhagen.

Quantum mechanics had won over many physicists by then, but it struck chemists as ugly and unintuitive. This wasn't stodginess as much as pragmatism: that funny way of counting electrons seemed to have little to do with real chemistry. However, Bohr's predictions about hafnium, made without setting foot in a lab, forced chemists to swallow hard. Coincidentally, Hevesy and Coster made their discovery just before Bohr accepted the 1922 Nobel Prize in Physics. They informed him by telegram in Stockholm, and Bohr announced their discovery in a speech. This made quantum mechanics look like the evolutionary science, since it dug deeper into atomic structure than chemistry could. A whispering campaign began, and as with Mendeleev before him, Bohr's colleagues soon imbued Bohr—already inclined to scientific mysticism—with oracular qualities.

That's the legend anyway. The truth is a little different. At least three scientists preceding Bohr, including a chemist who directly influenced Bohr, wrote papers as far back as 1895 that linked element seventy-two to transition metals such as zirconium. These men weren't geniuses ahead of their time, but pedestrian chemists with little knowledge of or interest in quantum physics. It seems that Bohr poached their arguments when placing hafnium and probably used his quantum calculations to rationalize a less romantic, but still viable, *chemical* argument about its spot on the table.*

Yet, as with most legends, what's important isn't the truth but the consequences—how people react to a story. And as the myth was bruited about, people clearly wanted to believe that Bohr had found hafnium through quantum mechanics alone. Physics had always worked by reducing nature's machines into smaller pieces, and for many scientists Bohr had reduced dusty, fusty chemistry to a specialized, and suddenly quaint, branch of physics. Philosophers of science also leapt on the story to proclaim that Mendeleevian chemistry was dead and Bohrian physics ruled the realm. What started as a scientific argument became a political dispute about territory and boundaries. Such is science, such is life.

The legend also lionized the man at the center of the brouhaha, György Hevesy. Colleagues had already nominated Hevesy for a Nobel Prize by 1924 for discovering hafnium, but there was a dispute over priority with a French chemist and dilettante painter. Georges Urbain—who had once tried and failed to embarrass Henry Moseley with his sample of rare earth elements—had discovered lutetium in 1907. Much later he claimed he had found hafnium—a rare earth flavor of hafnium—mixed in with his samples. Most scientists didn't find Urbain's work convincing, and unfortunately Europe was still divided by the recent unpleasantries in 1924, so the priority dispute took on nationalistic overtones. (The French considered Bohr and Hevesy Germans even though they were Danish and Hungarian, respectively. One French periodical sniffed that the whole thing "stinks of Huns," as if Attila himself had discovered the element.) Chemists also mistrusted Hevesy for his dual "citizenship" in chemistry and physics, and that, along with the political bickering, prevented the Nobel committee from giving him the prize. Instead, it left the 1924 prize blank.

Saddened but unbowed, Hevesy left Copenhagen for Germany and continued his important experiments on chemical tracers. In his spare time, he even helped determine how quickly the human body recycles an average water molecule (nine days) by volunteering to drink special "heavy" water,* in which some hydrogen atoms have an extra neutron, and then having his urine weighed each day. (As with the landlady-meat incident, he wasn't big on formal research protocol.) All the while, chemists such as Irène Joliot-Curie repeatedly and futilely nominated him for a Nobel Prize. Annually unrewarded, Hevesy grew a little despondent. But unlike with Gilbert Lewis, the obvious injustice aroused sympathy for Hevesy, and the lack of a prize strangely bolstered his status in the international community.

Nonetheless, with his Jewish ancestry, Hevesy soon faced direr problems than the lack of a Nobel Prize. He left Nazi Germany in the 1930s for Copenhagen again and remained there through August 1940, when Nazi storm troopers knocked on the front door of Bohr's institute. And when the hour called for it, Hevesy proved himself courageous. Two Germans, one Jewish and the other a Jewish sympathizer and defender, had sent their gold Nobel Prize medals to Bohr for safekeeping in the 1930s, since the Nazis would likely plunder them in Germany. However, Hitler had made exporting gold a state crime, so the discovery of the medals in Denmark could lead to multiple executions. Hevesy suggested they bury the medals, but Bohr thought that was too obvious. So, as Hevesy later recalled, "while the invading forces marched in the streets of Copenhagen, I was busy dissolving [Max von] Laue's and also James Franck's medals." To do this, he used aqua regia—a caustic mix of nitric and hydrochloric acids that fascinated alchemists because it dissolved "royal metals" such as gold (though not easily, Hevesy remembered). When the Nazis ransacked Bohr's institute, they scoured the building for loot or evidence of wrongdoing but left the beaker of orange aqua regia untouched. Hevesy was forced to flee to Stockholm in 1943, but when he returned to his battered laboratory after V-E Day, he found the innocuous beaker undisturbed on a shelf. He precipitated out the gold, and the Swedish Academy later recast the medals for Franck and Laue. Hevesy's only complaint about the ordeal was the day of lab work he missed while fleeing Copenhagen.

Amid those adventures, Hevesy continued to collaborate with colleagues, including Joliot-Curie. In fact, Hevesy was an unwitting witness to an enormous blunder by Joliot-Curie, which prevented her from making one of the great scientific discoveries of the twentieth century. That honor fell to another woman, an Austrian Jew, who, like Hevesy, fled Nazi persecution. Unfortunately, Lise Meitner's run-in with politics, both worldly and scientific, ended rather worse than Hevesy's.

Meitner and her slightly younger collaborator, Otto Hahn, began working together in Germany just before the discovery of element ninety-one. Its discoverer, Polish chemist Kazimierz Fajans, had detected only short-lived atoms of the element in 1913, so he named it "brevium." Meitner and Hahn realized in 1917 that most atoms of it actually live hundreds of thousands of years, which made "brevium" sound a little stupid. They rechristened it protactinium, or "parent of actinium," the element into which it (eventually) decayed.

No doubt Fajans protested this rejection of "brevium."

Although he was admired for his grace among high-society types, contemporaries say the Pole had a pugnacious and tactless streak in professional matters. Indeed, there's a myth that the Nobel committee had voted to award Fajans the vacant 1924 chemistry prize (the one Hevesy supposedly missed) for work on radioactivity but rescinded it, as punishment for hubris, when a photo of Fajans and a story headlined "K. Fajans to Receive Nobel Prize" appeared in a Swedish newspaper before the formal announcement. Fajans always maintained that an influential and unfriendly committee member had blocked him for personal reasons.* (Officially, the Swedish Academy said it had left that year's prize blank and kept the prize money to shore up its endowment, which, it complained, had been decimated by high Swedish taxes. But it released that excuse only after a public outcry. At first it just announced there would be no prizes in multiple categories and blamed "a lack of qualified candidates." We may never know the real story, since the academy says that "such information is deemed secret for all times.")

Regardless, "brevium" lost out, "protactinium" stuck,* and Meitner and Hahn sometimes receive credit for codiscovering element ninety-one today. However, there's another, more intriguing story to unpack in the work that led to the new name. The scientific paper that announced the long-lived protactinium betrayed the first signs of Meitner's unusual devotion to Hahn. It was nothing sexual—Meitner never married, and no one has ever found evidence she had a lover—but at least professionally, she was smitten with Hahn. That's probably because Hahn had recognized her worth and chosen to work alongside her in a retrofitted carpentry shop when German officials refused to give Meitner, a woman, a real lab. Isolated in the shop, they fell into a pleasing relationship where he performed the chemistry, identifying what elements were present in radioactive samples, and she performed the physics, figuring out how Hahn had gotten what he said. Unusually, though, Meitner performed *all* the work for the final, published protactinium experiments because Hahn was distracted with Germany's gas warfare during World War I. She nevertheless made sure he received credit. (Remember that favor.)

After the war, they resumed their partnership, but while the decades between the wars were thrilling in Germany scientifically, they proved scary politically. Hahn—strong-jawed, mustached, of good German stock—had nothing to fear from the Nazi takeover in 1932. Yet to his credit, when Hitler ran all the Jewish scientists out of the country in 1933—causing the first major wave of refugee scientists—Hahn resigned his professorship in protest (though he continued to attend seminars). Meitner, though raised a proper Austrian Protestant, had Jewish grandparents. Characteristically, and perhaps because she had at last earned her own real research lab, she downplayed the danger and buried herself in scintillating new discoveries in nuclear physics.

The biggest of those discoveries came in 1934, when Enrico Fermi announced that by pelting uranium atoms with atomic particles, he had fabricated the first transuranic elements. This wasn't true, but people were transfixed by the idea that the periodic table was no longer limited to ninety-two entries. A fireworks display of new ideas about nuclear physics kept scientists busy around the world.

That same year, another leader in the field, Irène Joliot-Curie, did her own bombardments. After careful chemical analysis, she announced that the new transuranic elements betrayed an uncanny similarity to lanthanum, the first rare earth. This, too, was unexpected—so unexpected that Hahn didn't believe it. Elements bigger than uranium simply could not behave exactly like a tiny metallic element nowhere near uranium on the periodic table. He politely told Frédéric Joliot-Curie that the lanthanum link was hogwash and vowed to redo Irène's experiments to show that the transuranics were nothing like lanthanum.

Also in 1938, Meitner's world collapsed. Hitler boldly annexed Austria and embraced all Austrians as his Aryan brethren — except anyone remotely Jewish. After years of willed invisibility, Meitner was suddenly subject to Nazi pogroms. And when a colleague, a chemist, tried to turn her in, she had no choice but to flee, with just her clothes and ten deutsch marks. She found refuge in Sweden and accepted a job at, ironically, one of the Nobel science institutes.

Despite the hardships, Hahn remained faithful to Meitner, and the two continued to collaborate, writing letters like clandestine lovers and occasionally rendezvousing in Copenhagen. During one such meeting in late 1938, Hahn arrived a little shaken. After repeating Irène Joliot-Curie's experiments, he had found her elements. And they not only behaved *like* lanthanum (and another nearby element she'd found, barium), but, according to every known chemical test, they *were* lanthanum and barium. Hahn was considered the best chemist in the world, but that finding "contradict[ed] all previous experience," he later admitted. He confessed his humbling bafflement to Meitner.

Meitner wasn't baffled. Out of all the great minds who worked on transuranic elements, only hard-eyed Meitner grasped that they weren't transuranic at all. She alone (after discussions with her nephew and new partner, physicist Otto Frisch) realized that Fermi hadn't discovered new elements; he'd discovered nuclear fission. He'd cracked uranium into smaller elements and misinterpreted his results. The eka-lanthanum

Joliot-Curie had found was plain lanthanum, the fallout of the first tiny nuclear explosions! Hevesy, who saw early drafts of Joliot-Curie's papers from that time, later reminisced on how close she'd come to making that unimaginable discovery. But Joliot-Curie, Hevesy said, "didn't trust herself enough" to believe the correct interpretation. Meitner trusted herself, and she convinced Hahn that everyone else was wrong.

Naturally, Hahn wanted to publish these astounding results, but his collaboration with, and debt to, Meitner made doing so politically tricky. They discussed options, and she, deferential, agreed to name just Hahn and his assistant on the key paper. Meitner and Frisch's theoretical contributions, which made sense of everything, appeared later in a separate journal. With those publications, nuclear fission was born just in time for Germany's invasion of Poland and the start of World War II.

So began an improbable sequence of events that culminated in the most egregious oversight in the history of the Nobel Prize. Even without knowledge of the Manhattan Project, the Nobel committee had decided by 1943 to reward nuclear fission with a prize. The question was, who deserved it? Hahn, clearly. But the war had isolated Sweden and made it impossible to interview scientists about Meitner's contributions, an integral part of the committee's decision. The committee therefore relied on scientific journals—which arrived months late or not at all, and many of which, especially prestigious German ones, had barred Meitner. The emerging divisions between chemistry and physics also made it hard to reward interdisciplinary work.

After suspending the prizes in 1940, the Swedish Academy began awarding a few retroactively in 1944. First up, at long last, Hevesy won the vacant 1943 prize for chemistry—though perhaps partly as a political gesture, to honor all refugee scientists.

In 1945, the committee took up the more vexed matter of fission. Meitner and Hahn both had strong back-room advocates on the Nobel committee, but Hahn's backer had the chutzpah to point out that Meitner had done no work "of great importance" in the previous few years—when she was in hiding from Hitler. (Why the committee never directly interviewed Meitner, who was working at a nearby Nobel institute, isn't clear, although it's generally bad practice to interview people about whether they deserve a prize.) Meitner's backer argued for a shared prize and probably would have prevailed given time. But when he died unexpectedly, the Axis-friendly committee members mobilized, and Hahn won the 1944 prize alone.

Shamefully, when Hahn got word of his win (the Allies now had him in military custody for suspicion of working on Germany's atomic bomb; he was later cleared), he didn't speak up for Meitner. As a result, the woman he'd once esteemed enough to defy his bosses and work with in a carpentry shop got nothing—a victim, as a few historians had it, of "disciplinary bias, political obtuseness, ignorance, and haste."*

The committee could have rectified this in 1946 or later, of course, after the historical record made Meitner's contributions clear. Even architects of the Manhattan Project admitted how much they owed her. But the Nobel committee, famous for what *Time* magazine once called its "old-maid peevishness," is not prone to admit mistakes. Despite being repeatedly nominated her whole life—by, among others, Kazimierz Fajans, who knew the pain of losing a Nobel better than anyone—she died in 1968 without her prize.

Happily, however, "history has its own balance sheet." The transuranic element 105 was originally named hahnium, after Otto Hahn, by Glenn Seaborg, Al Ghiorso, and others in 1970. But during the dispute over naming rights, an international committee—as if hahnium was Poland—stripped the element of that name in 1997, dubbing it dubnium. Because of the peculiar rules for naming elements*—basically, each name gets one shot—hahnium can never be considered as the name for a new element in the future, either. The Nobel Prize is all Hahn gets. And the committee soon crowned Meitner with a far more exclusive honor than a prize given out yearly. Element 109 is now and forever will be known as meitnerium.

13

Elements as Money

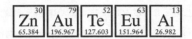

If the periodic table has a history with politics, it has an even longer and cozier relationship with money. The stories of many metallic elements cannot be told without getting tangled up in the history of money, which means the history of those elements is also tangled up with the history of counterfeiting. In different centuries, cattle, spices, porpoise teeth, salt, cocoa beans, cigarettes, beetle legs, and tulips have all passed for currency, none of which can be faked convincingly. Metals are easier to doctor. Transition metals especially have similar chemistries and densities because they have similar electron structures, and they can blend together and replace one another in alloys. Different combinations of precious and less-than-precious metals have been fooling people for millennia.

Around 700 BC, a prince named Midas inherited the kingdom of Phrygia in what is now Turkey. According to various myths (which might conflate two rulers named Midas), he led an eventful life. Jealous Apollo, the god of music, asked Midas to judge a showdown between him and the other great lyre strummers of the era, then turned Midas's ears into donkey ears when Midas favored someone else over Apollo. (He didn't deserve human ears if he judged music that badly.) Midas also reportedly maintained the best rose garden in the ancient world. Scientifically, Midas sometimes receives credit for discovering tin (not true, though it was mined in his kingdom) and for discovering the minerals "black lead" (graphite) and "white lead" (a beautiful, bright white, poisonous lead pigment). Of course, no one would remember Midas today if not for another metallurgical novelty, his golden touch. He earned it after tending to the drunken satyr Silenus, who passed out in his rose garden one night. Silenus so appreciated the monarch's hospitality that he offered Midas a reward. Midas asked that whatever he touched transform into gold—a delight that soon cost him his daughter when he hugged her and almost cost him his life, since for a time even food transubstantiated into gold at his lips.

Obviously, none of that probably ever happened to the real king. But there's evidence that Midas earned his legendary status for good reason. It all traces back to the Bronze Age, which began in Midas's neighborhood around 3000 BC. Casting bronze, an alloy of tin and copper, was the high-tech field of the day, and although the metal remained expensive, the technology had penetrated most kingdoms by the time of Midas's reign. The skeleton of a king popularly called Midas (but proved later to be his father, Gordias) was found in its tomb in Phrygia surrounded by bronze cauldrons and handsome bronze bowls with inscriptions, and the otherwise naked skeleton itself was found wearing a bronze belt. But in saying "bronze," we need to be more specific. It's not like water, where two parts hydrogen always combine with one part oxygen. A number of different alloys with different ratios of metals all count as bronze, and bronze metals around the ancient world differed in color depending on the percentages of tin, copper, and other elements where the metals were mined.

One unique feature of the metallic deposits near Phrygia was the abundance of ores with zinc. Zinc and tin ores commonly commingle in nature, and deposits of one metal can easily be mistaken for the other. What's interesting is that zinc mixed with copper doesn't form bronze; it forms brass. And the earliest known brass foundries existed in, of all places, the part of Asia Minor where Midas once reigned.

Is it obvious yet? Go find something bronze and something brass and examine them. The bronze is shiny, but with overtones of copper. You wouldn't mistake it for anything else. The shine of brass is more alluring, more subtle, a little more... golden. Midas's touch, then, was possibly nothing more than an accidental touch of zinc in the soil of his corner of Asia Minor.

To test that theory, in 2007 a professor of metallurgy at Ankara University in Turkey and a few historians constructed a primitive Midas-era furnace, into which they loaded local ores. They melted them, poured the resulting liquid into molds, and let it cool. Mirabile dictu, it hardened into an uncannily golden bullion. Naturally, it's impossible to know whether the contemporaries of King Midas believed that his precious zinc-laden bowls and statues and belts were actually gold. But they weren't necessarily the ones making up legends about him. More probably, the Greek travelers who later colonized that region of Asia Minor simply grew smitten with the Phrygian "bronzes," so much brighter than their own. The tales they sent home could have swelled century by century, until golden-hued brass transmuted into real gold, and a local hero's earthly power transmuted into the supernatural power to create precious metals at a touch. After that, it took only the genius of Ovid to touch up the story for his *Metamorphoses*, and voilà: a myth with a more-than-plausible origin.

An even deeper archetype in human culture than Midas is the lost city of gold — of travelers in far-off, alien lands stumbling onto unimaginable wealth. El Dorado. In modern and (slightly) more realistic times, this dream often takes the form of gold rushes. Anyone who paid an iota of attention in history class knows that real gold rushes were awful, dirty, dangerous affairs, with bears and lice and mine collapses and lots of pathetic whoring and gambling. And the chances that a person would end up rich were almost zilch. Yet almost no one with an iota of imagination hasn't dreamed of throwing over everything in his humdrum life and rushing off to prospect for a few pure nuggets. The desire for a great adventure and the love of riches are practically built into human nature. As such, history is dotted with innumerable gold rushes.

Nature, naturally, doesn't want to part with her treasure so easily, so she invented iron pyrite (iron disulfide) to thwart amateur prospectors. Perversely, iron pyrite shines with a luster more golden than real gold, like cartoon gold or gold in the imagination. And more than a few greenhorns and people blinded by greed have been taken in during a fool's gold rush. But in all of history, probably the most confounded gold rush ever took place in 1896, on rough frontier land in the Australian outback. If iron pyrite is faux gold, then this gold rush in Australia—which eventually found desperate prospectors knocking down their own chimneys with pickaxes and sifting through the rubble—was perhaps the first stampede in history caused by "fool's fool's gold."

Three Irishmen, including Patrick (Paddy) Hannan, were traversing the outback in 1893 when one of their horses lost a shoe twenty miles from home. It might have been the luckiest breakdown in history. Within days, without having to dig an inch into the dirt, they'd collected eight pounds of gold nuggets just walking around. Honest but dim, the trio filed a claim with territory officials, which put the location on public record.

Within a week, hundreds of prospectors were storming Hannan's Find, as the post became known, to try their luck.

In a way, the expanse was easy pickings. During those first months in the desert, gold was more plentiful than water. But while that sounds great, it wasn't. You can't drink gold, and as more and more miners piled in, the prices of supplies soared, and competition for mining sites grew fierce. People started having to dig for gold, and some fellows figured out there was easier money to be had in building up a real town instead. Breweries and brothels popped up in Hannan's Find, as did houses and even paved roads. For bricks, cement, and mortar, builders collected the excess rock dug out during excavations. Miners just cast it aside, and as long as they were going to keep digging, there was nothing better to do with the rubble.

Or so they assumed. Gold is an aloof metal. You won't find it mixed inside minerals and ores, because it doesn't bond with other elements. Its flakes and nuggets are usually pure, besides a few odd alloys. The exception, the single element that will bond to gold, is tellurium, a vampirish element first isolated in Transylvania in 1782. Tellurium combines with gold to form some garish-sounding minerals—krennerite, petzite, sylvanite, and calaverite—with some equally atrocious chemical formulas. Instead of nice proportions such as H_2O and CO_2, krennerite is $(Au_{0.8}, Ag_{0.2})Te_2$. Those tellurides vary in color, too, and one of them, calaverite, shines sort of yellow.

Actually it shines more like brass or iron pyrite than deep-hued gold, but it's probably close enough to trick you if you've been out in the sun all day. You can imagine a raw, dirty eighteen-year-old hauling in calaverite nuggets to the local appraiser in Hannan's Find, only to hear the appraiser dismiss them as a sackful of what mineralogists classify as bagoshite. Remember, too, that some tellurium compounds (not calaverite, but others) smell pungent, like garlic magnified a thousand times, an odor notoriously difficult to get rid of. Better to sell it and bury it in roads, where it won't stink, and get back to digging for the real McCoy.

But people just kept piling into Hannan's Find, and food and water didn't get any cheaper. At one point, tensions over supplies ran so high that a full-on riot erupted. And as things got desperate, rumors circulated about that yellowish tellurium rock they were digging up and throwing away. Even if hardscrabble miners weren't acquainted with calaverite, geologists had been for years and knew its properties. For one, it decomposes at low temperatures, which makes separating out the gold darn easy. Calaverite had first been found in Colorado in the 1860s.* Historians suspect that campers who'd built a fire one night noticed that, um, the rocks they'd ringed the fire pit with were oozing gold. Pretty soon, stories to that effect made their way to Hannan's Find.

Hell finally broke loose on May 29, 1896. Some of the calaverite used to build Hannan's Find contained five hundred ounces of gold per ton of rock, and miners were soon tearing out every damn ounce they could find. People attacked refuse heaps first, scrabbling among them for discarded rock. When those were picked clean, they went after the town itself. Paved-over potholes became potholes again; sidewalks were chiseled out; and you can bet the miner who built the chimney and hearth for his new home out of gold telluride–infused brick wasn't too sentimental about tearing it apart again.

In later decades, the region around Hannan's Find, soon renamed Kalgoorlie, became the world's largest gold producer. The Golden Mile, they called it, and Kalgoorlie bragged that its engineers outpaced the rest of the world when it came to extracting gold from the ground. Seems like the later generations learned their lesson—not to be throwing aside rocks all willy-nilly—after their fathers' fool's fool's gold rush.

Midas's zinc and Kalgoorlie's tellurium were rare cases of unintentional deception: two innocent moments in monetary history surrounded by aeons of deliberate counterfeiting. A century after Midas, the first real money, coins made of a natural silver-gold alloy called electrum, appeared in Lydia, in Asia Minor. Shortly after that, another fabulously wealthy ancient ruler, the Lydian king Croesus, figured out how to separate electrum into silver and gold coins, in the process establishing a real currency system. And within a few years of Croesus's feat, in 540 BC, King Polycrates, on the Greek isle Samos, began buying off his enemies in Sparta with lead slugs plated with gold. Ever since then, counterfeiters have used elements such as lead, copper, tin, and iron the way cheap barkeeps use water in kegs of beer—to make the real money stretch a little further.

Today counterfeiting is considered a straight case of fraud, but for most of history, a kingdom's precious-metal currency was so tied up with its economic health that kings considered counterfeiting a high crime — treason. Those convicted of such treason faced hanging, if not worse. Counterfeiting has always attracted people who do not understand opportunity costs— the basic economic law that you can make far more money plying an honest trade than spending hundreds of hours making "free" money. Nevertheless, it has taken some brilliant minds to thwart those criminals and design anything approaching foolproof currency.

For instance, long after Isaac Newton had derived the laws of calculus and his monumental theory of gravity, he became master of the Royal Mint of England in the last few years of the 1600s. Newton, in his early fifties, just wanted a well-paying government post, but to his credit he didn't approach it as a sinecure. Counterfeiting—especially "clipping" coins by shaving the edges and melting the scraps together to make new coins—was endemic in the seedier parts of London. The great Newton found himself entangled with spies, lowlifes, drunkards, and thieves—an entanglement he thoroughly enjoyed. A pious Christian, Newton prosecuted the wrongdoers he uncovered with the wrath of the Old Testament God, refusing pleas for clemency. He even had one notorious but slippery "coiner," William Chaloner—who'd goaded Newton for years with accusations of fraud at the mint—hanged and publicly disemboweled.

The counterfeiting of coins dominated Newton's tenure, but not long after he resigned, the world financial system faced new threats from fake paper currency. A Mongol emperor in China, Kublai Khan, had introduced paper money there in the 1200s. The innovation spread quickly in Asia at first—partly because Kublai Khan executed anyone who refused to use it—but only intermittently in Europe. Still, by the time the Bank of England began issuing paper notes in 1694, the advantages of paper currency had grown obvious. The ores for making coins were expensive, coins themselves were cumbersome, and the wealth based on them depended too much on unevenly distributed mineral resources. Coins also, since knowledge of metalworking was more widespread in centuries past, were easier for most people to counterfeit than paper money. (Nowadays the situation is vice versa. Anyone with a laser printer can make a decent $20 bill. Do you have a single acquaintance who could cast a passable nickel, even if such a thing were worth doing?)

If the alloy-friendly chemistry of metal coins once favored swindlers, in the age of paper money the unique chemistry of metals like europium helps governments combat swindling.

It all traces back to the chemistry of europium, especially the movement of electrons within its atoms. So far we've discussed only electron bonds, the movement of electrons between atoms. But electrons constantly whirl around their own nuclei, too, movement often compared to planets circling a sun. Although that's a pretty good analogy, it has a flaw if taken literally. Earth in theory could have ended up on many different orbits around the sun. Electrons cannot take any old path around a nucleus. They move within shells at different energy levels, and because there's no energy level between one and two, or two and three, and so on, the paths of electrons are highly circumscribed: they orbit only at certain distances from their "sun" and orbit in oblong shapes at funny angles. Also unlike a planet, an electron—if excited by heat or light—can leap from its low-energy shell to an empty, high-energy shell. The electron cannot stay in the high-energy state for long, so it soon crashes back down. But this isn't a simple back-and-forth motion, because as it crashes, the electron jettisons energy by emitting light.

The color of the emitted light depends on the relative heights of the starting and ending energy levels. A crash between closely spaced levels (such as two and one) releases a pulse of low-energy reddish light, while a crash between more widely spaced levels (say, five and two) releases high-energy purple light. Because the electrons' options about where to jump are limited to whole-number energy levels, the emitted light is also constrained. Light emitted by electrons in atoms is not like the white light from a lightbulb. Instead, electrons emit light of very specific, very pure colors. Each element's shells sit at different heights, so each element releases characteristic bands of color—the very bands Robert Bunsen observed with his burner and spectroscope. Later, the realization that electrons jump to whole-number levels and never orbit at fractional levels was a fundamental insight of quantum mechanics. Everything wacky you've ever heard about quantum mechanics derives directly or indirectly from these discontinuous leaps.

Europium can emit light as described above, but not very well: it and its brother lanthanides don't absorb incoming light or heat efficiently (another reason chemists had trouble identifying them for so long). But light is an international currency, redeemable in many forms in the atomic world, and lanthanides can emit light in a way other than simple absorption. It's called fluorescence,* which is familiar to most people from black lights and psychedelic posters. In general, normal emissions of light involve just electrons, but fluorescence involves whole molecules. And whereas electrons absorb and emit light of the same color (yellow in, yellow out), fluorescent molecules absorb high-energy light (ultraviolet light) but emit that energy as lower-energy, visible light. Depending on the molecule it's attached to, europium can emit red, green, or blue light.

That versatility is a bugbear for counterfeiters and makes europium a great anticounterfeiting tool. The European Union (EU), in fact, uses its eponymous element in the ink on its paper bills. To prepare the ink, EU treasury chemists lace a fluorescing dye with europium ions, which latch onto one end of the dye molecules. (No one really knows which dyes, since the EU has reportedly outlawed looking into it. Law-abiding chemists can only guess.) Despite the anonymity, chemists know the europium dyes consist of two parts. First is the receiver, or antenna, which forms the bulk of the molecule. The antenna captures incoming light energy, which europium cannot absorb; transforms it into vibrational energy, which europium can absorb; and wriggles that energy down to the molecule's tip. There the europium stirs up its electrons, which jump to higher energy levels. But just before the electrons jump and crash and emit, a bit of the incoming wave of energy "bounces" back into the antenna. That wouldn't happen with isolated europium atoms, but here the bulky part of the molecule dampens the energy and dissipates it. Because of that loss, when electrons crash back down, they produce lower-energy light.

So why is that shift useful? The fluorescing dyes are selected so that europium appears dull under visible light, and a counterfeiter might be lulled into thinking he has a perfect replica. Slide a euro note beneath a special laser, though, and the laser will tickle the invisible ink. The paper itself goes black, but small, randomly oriented fibers laced with europium pop out like parti-colored constellations. The charcoal sketch of Europe on the bills glows green, as it might look to alien eyes from space. A pastel wreath of stars gains a corona of yellow or red, and monuments and signatures and hidden seals shine royal blue. Officials nab counterfeits simply by looking for bills that don't show all these signs.

There are really two euros on each banknote, then: the one we see day to day and a second, hidden euro mapped directly onto the first—an embedded code. This effect is extremely hard to fake without professional training, and the europium dyes, in tandem with other security features, make the euro the most sophisticated piece of currency ever devised. Euro banknotes certainly haven't stopped counterfeiting; that's probably impossible as long as people like holding cash. But in the periodic table–wide struggle to slow it down, europium has taken a place among the most precious metals.

Despite all the counterfeiting, many elements have been used as legitimate currency throughout history. Some, such as antimony, were a bust. Others became money under gruesome circumstances. While working in a prison chemical plant during

World War II, the Italian writer and chemist Primo Levi began stealing small sticks of cerium. Cerium sparks when struck, making it an ideal flint for cigarette lighters, and he traded the sticks to civilian workers in exchange for bread and soup. Levi came into the concentration camps fairly late, nearly starved there, and began bartering with cerium only in November 1944. He estimated that it bought him two months' worth of rations, of life, enough to last until the Soviet army liberated his camp in January 1945. His knowledge of cerium is why we have his post-Holocaust masterpiece *The Periodic Table* today.

Other proposals for elemental currency were less pragmatic and more eccentric. Glenn Seaborg, caught up in nuclear enthusiasm, once suggested that plutonium would become the new gold in world finance, because it's so valuable for nuclear applications. Perhaps as a send-up of Seaborg, a science fiction writer once suggested that radioactive waste would be a better currency for global capitalism, since coins stamped from it would certainly circulate quickly. And, of course, during every economic crisis, people bellyache about reverting to a gold or silver standard. Most countries considered paper bills the equivalent of actual gold or silver until the twentieth century, and people could freely trade the paper for the metal. Some literary scholars think that L. Frank Baum's 1900 book *The Wonderful Wizard of Oz*—whose Dorothy wore silver, not ruby, slippers and traveled on a gold-colored brick road to a cash-green city—was really an allegory about the relative merits of the silver versus the gold standard.

However antiquated a metals-based economy seems, such people have a point. Although metals are quite illiquid, the metals markets are one of the most stable long-term sources of wealth. It doesn't even have to be gold or silver. Ounce by ounce, the most valuable element, among the elements you can actually buy, is rhodium. (That's why, to trump a mere platinum record, the *Guinness Book of Records* gave former Beatle Paul McCartney a disk made of rhodium in 1979 to celebrate his becoming the bestselling musician of all time.) But no one ever made more money more quickly with an element on the periodic table than the American chemist Charles Hall did with aluminium.

A number of brilliant chemists devoted their careers to aluminium throughout the 1800s, and it's hard to judge whether the element was better or worse off afterward. A Danish chemist and a German chemist simultaneously extracted this metal from the ancient astringent alum around 1825. (Alum is the powder cartoon characters like Sylvester the cat sometimes swallow that makes their mouths pucker.) Because of its luster, mineralogists immediately classified aluminium as a precious metal, like silver or platinum, worth hundreds of dollars an ounce.

Twenty years later, a Frenchman figured out how to scale up these methods for industry, making aluminium available commercially. For a price. It was still more expensive than even gold. That's because, despite being the most common metal in the earth's crust—around 8 percent of it by weight, hundreds of millions of times more common than gold—aluminium never appears in pure, mother lode-al form. It's always bonded to something, usually oxygen. Pure samples were considered miracles. The French once displayed Fort Knox–like aluminium bars next to their crown jewels, and the minor emperor Napoleon III reserved a prized set of aluminium cutlery for special guests at banquets. (Less favored guests used gold knives and forks.) In the United States, government engineers, to show off their country's industrial prowess, capped the Washington Monument with a six-pound pyramid of aluminium in 1884.

Dapper engineers refurbish the aluminium cap atop the Washington Monument. The U.S. government crowned the monument with aluminium in 1884 because it was the most expensive (and therefore most impressive) metal in the world, far dearer than gold. (Bettmann/Corbis)

A historian reports that one ounce of shavings from the pyramid would have paid a day's wages for each of the laborers who erected it.

Aluminium's sixty-year reign as the world's most precious substance was glorious, but soon an American chemist ruined everything. The metal's properties—light, strong, attractive—tantalized manufacturers, and its omnipresence in the earth's crust had the potential to revolutionize metal production. It obsessed people, but no one could figure out an efficient way to separate it from oxygen. At Oberlin College in Ohio, a chemistry professor named Frank Fanning Jewett would regale his students with tales of the aluminium El Dorado that awaited whoever mastered this element. And at least one of his students had the naïveté to take his professor seriously.

In his later years, Professor Jewett bragged to old college chums that "my greatest discovery was the discovery of a man"—Charles Hall. Hall worked with Jewett on separating aluminium throughout his undergraduate years at Oberlin. He failed and failed and failed again, but failed a little more smartly each time. Finally, in 1886, Hall ran an electric current from handmade batteries (power lines didn't exist) through a liquid with dissolved aluminium compounds. The energy from the current zapped and liberated the pure metal, which collected in minute silver nuggets on the bottom of the tank. The process was cheap and easy, and it would work just as well in huge vats as on the lab bench. This had been the most sought-after chemical prize since the philosopher's stone, and Hall had found it. The "aluminium boy wonder" was just twenty-three.

Hall's fortune, however, was not made instantly. Chemist Paul Héroult in France stumbled on more or less the same process at the same time. (Today Hall and Héroult share credit for the discovery that crashed the aluminium market.) An Austrian invented another separation method in 1887, and with the competition bearing down on Hall, he quickly founded what became the Aluminum Company of America, or Alcoa, in Pittsburgh. It turned into one of the most successful business ventures in history.

Aluminium production at Alcoa grew at exponential rates. In its first months in 1888, Alcoa eked out 50 pounds of aluminium per day; two decades later, it had to ship 88,000 pounds per day to meet the demand. And while production soared, prices plummeted. Years before Hall was born, one man's breakthrough had dropped aluminium from $550 per pound to $18 per pound in seven years. Fifty years later, not even adjusting for inflation, Hall's company drove down the price to 25 cents per pound. Such growth has been surpassed probably only one other time in American history, during the silicon semiconductor revolution eighty years later,* and like our latter-day computer barons, Hall cleaned up. At his death in 1914, he owned Alcoa shares worth $30 million* (around $650 million today). And thanks to Hall, aluminium became the utterly blasé metal we all know, the basis for pop cans and pinging Little League bats and airplane bodies. (A little anachronistically, it still sits atop the Washington Monument, too.) I suppose it depends on your taste and temperament whether you think aluminium was better off as the world's most precious or most productive metal.

Incidentally, I use the international spelling "aluminium" instead of the strictly American "aluminum" throughout this book. This spelling disagreement* traces its roots back to the rapid rise of this metal. When chemists in the early 1800s speculated about the existence of element thirteen, they used both spellings but eventually settled on the extra *i*. That spelling paralleled the recently discovered barium, magnesium, sodium, and strontium. When Charles Hall applied for patents on his electric current process, he used the extra *i*, too. However, when advertising his shiny metal, Hall was looser with his language. There's debate about whether cutting the *i* was intentional or a fortuitous mistake on advertising fliers, but when Hall saw "aluminum," he thought it a brilliant coinage. He dropped the vowel permanently, and with it a syllable, which aligned his product with classy platinum. His new metal caught on so quickly and grew so economically important that "aluminum" became indelibly stamped on the American psyche. As always in the United States, money talks.

14

Artistic Elements

$\mathbf{A}$s science grew more sophisticated throughout its history, it grew correspondingly expensive, and money, big money, began to dictate if, when, and how science got done. Already by 1956, the German-English novelist Sybille Bedford could write* that many generations had passed since "the laws of the universe were something a man might deal with pleasantly in a workshop set up behind the stables."

Of course, very few people, mostly landed gentlemen, could have afforded a little workshop in which to do their science during the eras Bedford was pining for, the eighteenth and nineteenth centuries. To be sure, it's no coincidence that people from the upper classes were usually the ones doing things like discovering new elements: no one else had the leisure to sit around and argue about what some obscure rocks were made of.

This mark of aristocracy lingers on the periodic table, an influence you can read without an iota of knowledge about chemistry. Gentlemen throughout Europe received educations heavy in the classics, and many element names—cerium, thorium, promethium—point to ancient myths. The really funny-looking names, too, such as praseodymium, molybdenum, and dysprosium, are amalgams of Latin and Greek. Dysprosium means "little hidden one," since it's tricky to separate from its brother elements. Praseodymium means "green twin" for similar reasons (its other half is neodymium, "new twin"). The names of noble gases mostly mean "stranger" or "inactive." Even proud French gentlemen as late as the 1880s chose not "France" and "Paris" when enshrining new elements but the philologically moribund "Gallia" (gallium) and "Lutetia" (lutetium), respectively, as if sucking up to Julius Caesar.

All this seems odd today—scientists receiving more training in antique languages than, well, in science—but for centuries science was less a profession than a hobby* for amateurs, like philately. Science wasn't yet mathematized, the barriers for entry were low, and a nobleman with the clout of, say, Johann Wolfgang von Goethe could bully his way into scientific discussions, qualified or not.

Today Goethe is remembered as a writer whose range and emotive power many critics rank second only to Shakespeare's, and beyond his writing, he took an active role in government and in policy debates in nearly every field. Many people still rank him as the greatest, most accomplished German ever to live. But I have to admit that my first impression of Goethe was that he was a bit of a fraud.

One summer in college, I worked for a physics professor who, though a wonderful storyteller, was forever running out of really basic supplies like electronic cables, which meant I had to visit the departmental supply room in the basement to beg. The dungeon master there was a German-speaking man. In keeping with his Quasimodo-like job, he was often unshaven and had shoulder-length, tendriled hair, and his big arms and chest would have seemed hulking had he stood taller than five feet six. I trembled every time I knocked on his door, never knowing what to sputter back when he narrowed his eyes and said, more a scoff than a question, "He duzzno have any cohackzial cable?"

My relationship with him improved the next semester when I took a (required) course he co-taught. It was a lab, which meant tedious hours building and wiring things, and during the dead time he and I talked literature once or twice. One day he mentioned Goethe, whom I didn't know. "He's the Shakezpeare of Germany," he explained. "All the stuck-up German azzholes, all the time they quote him. It's dizgusting. Then they say, 'What, you don't know *Goethe*?'"

He had read Goethe in the original German and found him mediocre. I was still young enough to be impressed by any strong convictions, and the denunciation made me suspicious of Goethe as a great thinker. Years later, after reading more widely, I came to appreciate Goethe's literary talent. But I had to admit my lab director had a point about Goethe's mediocrity in some areas. Though an epochal, world-changing author, Goethe couldn't hold back from making pronouncements in philosophy and science, too. He did so with all the enthusiasm of a dilettante, and about as much competence.

In the late 1700s, Goethe devised a theory of how colors work, to refute Isaac Newton's theory; except Goethe's relied as much on poetry as science, including his whimsical thesis that "colors are the deeds of light, deeds and sufferings." Not to huff like a positivist, but that statement has absolutely no meaning. He also laded his novel *Elective Affinities* with the spurious idea that marriages work like chemical reactions. That is, if you throw couple AB into contact with couple CD, they all might naturally commit chemical adultery and form new pairs: $AB + CD \rightarrow AD + BC$. And this wasn't just implied or a metaphor. Characters actually discuss this algebraic rearrangement of their lives. Whatever the novel's other strengths (especially its depiction of passion), Goethe would have been better off cutting out the science.

Even Goethe's masterwork, *Faust*, contains hoary speculation on alchemy and, worse (alchemy is at least cool), a bootless Socratic dialogue between "Neptunists" and "Plutonists"* on how rocks form. Neptunists like Goethe thought rocks precipitated from minerals in the ocean, the realm of the god Neptune; they were wrong. Plutonists—who were named after the underworld god Pluto and whose argument was taken up, in a rather unsubtle dig, by Satan himself in *Faust*—argued correctly that volcanoes and heat deep within the earth form most rocks. As usual, Goethe picked the losing side because it pleased him aesthetically. *Faust* remains as powerful a tale of scientific hubris as *Frankenstein*, but Goethe would have been crushed after his death in 1832 to learn that its science and philosophy would soon disintegrate and that people now read his work strictly for its literary value.

Nevertheless, Goethe did make one lasting contribution to science generally and the periodic table specifically—through patronage. In 1809, as a minister of the state, Goethe had the responsibility to pick a scientist for an open chair in chemistry at the University of Jena. After hearing recommendations from friends, Goethe had the foresight to select another Johann Wolfgang—J. W. Döbereiner. He was a provincial man with no chemistry degree and a poor résumé, having given chemistry a shot only after failing in the drug, textile, agricultural, and brewing industries. Döbereiner's work in industry, however, taught him practical skills that a gentleman like Goethe never learned but much admired during an age of great industrial leaps. Goethe soon developed a strong interest in the young man, and they spent many happy hours discussing hot chemistry topics of the day, such as why red cabbage tarnishes silver spoons and what the ingredients in Madame de Pompadour's toothpaste were. But the friendship couldn't quite erase vast differences in background and education. Goethe, naturally, had received a broadly classical education, and even today he is often hailed (with a touch of hyperbole) as the last man who knew everything, which was still possible back when art, science, and philosophy overlapped a great deal. He was also a much-traveled cosmopolitan. When Goethe tapped him for the post in Jena, Döbereiner had never even left Germany before, and gentlemen intellects like Goethe remained far more typical scientists than bumpkins like the lesser J.W.

It's fitting, then, that Döbereiner's greatest contribution to science was inspired by one of the rare elements, strontium, whose name is neither Hellenic nor based on something in Ovid. Strontium was the first flicker that something like the periodic table existed. A doctor discovered it in a hospital lab in London's red-light district in 1790, not far from Shakespeare's old Globe Theatre. He named it after the origin of the minerals he was studying—Strontian, a mining village in Scotland—and Döbereiner picked up his work twenty years later. Döbereiner's research focused (notice the practicality) on finding precise ways to weigh elements, and strontium was new and rare, a challenge. With Goethe's encouragement, he set out to study its characteristics. As he refined his figures on strontium, though, he noticed something queer: its weight fell exactly between the weights of calcium and barium. Moreover, when he looked into the chemistry of strontium, it behaved like barium and calcium in chemical reactions. Strontium was somehow a blend of two elements, one lighter and one heavier.

Intrigued, Döbereiner began to precisely weigh more elements, scouting around for other "triads." Up popped chlorine, bromine, and iodine; sulfur, selenium, and tellurium; and more. In each case, the weight of the middle element fell halfway between its chemical cousins. Convinced this was not a co-incidence, Döbereiner began to group these elements into what today we'd recognize as columns of the periodic table. Indeed, the chemists who erected the first periodic tables fifty years later started with Döbereiner's pillars.*

Now, the reason fifty years passed between Döbereiner and Dmitri Mendeleev without a periodic table was that the triad work got out of hand. Instead of using strontium and its neighbors to search for a universal way to organize matter, chemists (influenced by Christianity, alchemy, and the Pythagorean belief that numbers somehow embody true metaphysical reality) began seeing trinities everywhere and delving into triadic numerology. They calculated trilogies for the sake of calculating trilogies and elevated every three-in-one relationship, no matter how tenuous, into something sacred. Nevertheless, thanks to Döbereiner, strontium was the first element correctly placed in a larger universal scheme of elements. And Döbereiner never would have figured all this out without first the faith and then the support of Goethe.

Then again, Döbereiner did make his patron look like even more of a genius for supporting him all along when, in 1823, he invented the first portable lighter. This lighter relied on the curious ability of platinum to absorb and store massive amounts of burnable hydrogen gas. In an era when all cooking and heating still required fire, it proved an unfathomable economic boon. The lighter, called Döbereiner's lamp, actually made Döbereiner almost as famous worldwide as Goethe.

So even if Goethe made a poor show of things in his own scientific work, his writing helped spread the idea that science was noble, and his patronage nudged chemists toward the periodic table. He deserves at least an honorary position in the history of science—which, in the end, might have satisfied him. To quote no less a personage than Johann Wolfgang von Goethe (apologies to my old lab director!), "The history of science is science itself."

Goethe valued the intellectual beauty of science, and people who value beauty in science tend to revel in the symmetries of the periodic table and its Bach-like repetitions with variation. Yet not all the table's beauty is abstract. The table inspires art in all guises. Gold and silver and platinum are themselves lovely, and other elements, such as cadmium and bismuth, bloom into bright, colorful pigments in minerals or oil paints. Elements play a strong role in design, too, in the making of beautiful everyday objects. New alloys of elements often provide some subtle edge in strength or flexibility that transforms a design from functional to phenomenal. And with an infusion of the right element, something as humble as a fountain pen can achieve a design that—if it's not too embarrassing to say it (and for some pen aficionados, it's not)—inches close to majesty.*

In the late 1920s, the legendary Hungarian (and later American) designer László Moholy-Nagy drew an academic distinction between "forced obsolescence" and "artificial obsolescence." Forced obsolescence is the normal course of things for technologies, the roughage of history books: plows gave way to reapers, muskets to Gatling guns, wooden boat hulls to steel. In contrast, artificial obsolescence did and increasingly would dominate the twentieth century, Moholy-Nagy argued. People were abandoning consumer goods not because the goods were superannuated, but because the Joneses had some newer, fancier design. Moholy-Nagy—an artist and something of a philosopher of design—couched artificial obsolescence as materialistic, infantile, and a "moral disintegration." And as hard as it is to believe, the humble pen once seemed an example of people's gluttonous need for something, anything, advanced and all too *now*.

The career of the pen as Frodo's ring began in 1923 with one man. At twenty-eight, Kenneth Parker convinced the directors of the family business to concentrate the firm's money in a new design, his luxury Duofold pen. (He smartly waited until *Mr.* Parker, his dad, the big boss, had left for a long sea voyage around Africa and Asia and couldn't veto him.) Ten years later, in the worst days of the Great Depression, Parker gambled again by introducing another high-end model, the Vacumatic. And just a few years after that, Parker, by then boss himself, was itching for another new design. He had read and absorbed Moholy-Nagy's theories of design, but instead of letting the moral reproach of artificial obsolescence hem him in, Parker saw it in true American fashion: a chance to make a lot of money. If people had something better to buy, they would, even if they didn't need it. To this end, in 1941 he introduced what's widely considered the greatest pen in history, the Parker 51, named after the number of years the Parker Pen Company had been operating when this wonderful and utterly superfluous model hit the stores.

It was elegance herself. The pen's caps were gold- or chrome-plated, with a gold-feathered arrow for the pen's clasp. The body was as plump and tempting to pick up as a cigarillo and came in dandy colors such as Blue Cedar, Nassau Green, Cocoa, Plum, and Rage Red. The pen's head, colored India Black, looked like a shy turtle's head, which tapered to a handsome, calligraphic-style mouth. And from that mouth extended a tiny gold nib, like a rolled-up tongue, to dispense ink. Inside that sleek frame, the pen ran on a newly patented plastic called Lucite and a newly patented cylindrical system for delivering a newly patented

Aficionados often cite the Parker 51 as the greatest pen in history—as well as one of the most stylish designs ever, in any field. The pen's tip was fashioned from the rare and durable element ruthenium. (Jim Mamoulides, www.PenHero.com)

ink—ink that for the first time in penmanship history dried not by evaporation, while sitting *on* the paper, but by penetrating *into* the paper's fibers, drying via absorption in an instant. Even the way the cap snapped onto the pen body received two patents. Parker's engineers were scribal geniuses.

The only mole on this beauty was the tip of the gold nib, the part that actually touched the paper. Gold, a soft metal, deforms under the rigorous friction of writing. Parker originally capped the nib with a ring of osmiridium, an alloy of iridium and osmium. The two metals were suitably tough but scarce, expensive, and a headache to import. A sudden shortage or price hike might doom the design. So Parker hired a metallurgist away from Yale University to find a replacement. Within a year, the company filed for another patent for a ruthenium tip, an element little better than scrap until then. But it was a tip, finally, worthy of the rest of the design, and ruthenium began capping every Parker 51 in 1944.*

Now, honestly, despite its superior engineering, the Parker 51 was probably about equal to most pens in its basic job—delivering ink onto paper. But as design prophet Moholy-Nagy could have predicted, fashion trumped need. With its new tip, the company convinced consumers, through advertising, that human writing instruments had reached their apotheosis, and people began throwing away earlier Parker models to grab this one. The 51—"the world's most wanted pen"—became a status symbol, the only thing the classiest bankers, brokers, and politicians would sign checks, bar tabs, and golf scorecards with. Even Generals Dwight D. Eisenhower and Douglas MacArthur used 51s to sign the treaties that ended World War II in Europe and the Pacific in 1945. With such publicity, and with the optimism that washed over the world at the end of the war, sales jumped from 440,000 units in 1944 to 2.1 million in 1947—an amazing feat considering that the 51 cost at least $12.50 at the time and ran up to $50 ($100 to $400 today), and that the refillable ink cartridge and durable ruthenium tip meant that no one had to replace the pen.

Even Moholy-Nagy, though likely distressed at how smoothly his theories had translated into marketing, had to whistle at the 51. Its balance in the hand, its look, its creamy delivery of ink—Moholy-Nagy swooned, and once cited it as *the* perfect design. He even took a job consulting for Parker starting in 1944. After that, rumors persisted for decades that Moholy-Nagy had designed the 51 himself. Parker continued to sell various models of the 51 through 1972, and though twice as expensive as its next-cheapest competitor, it had outsold every pen ever made up to that point, reaping $400 million in sales (a few billion dollars today).

Of course, not long after the Parker 51 disappeared, the market for high-end pens began to shrivel. The reason is pretty obvious: while the 51 had thrived on making other pens seem inferior, pens were gradually being forced into obsolescence by technologies like the typewriter. But there's an ironic story to unbury in that takeover, which starts with Mark Twain and wends its way back to the periodic table.

After seeing a demonstration of a typewriter in 1874, and despite a worldwide economic depression, Twain ran right out and bought one for the outrageous price of $125 ($2,400 today). Within a week, he was writing letters on it (in all capitals; it had no lowercase) about how he looked forward to giving it away: "IT IS MOST TOO TEARING ON THE MIND," he lamented. It's sometimes hard to separate Twain's real complaints from his curmudgeonly persona, so maybe he was exaggerating. But by 1875, he had given away his typewriter and decided instead to endorse new "fountain" pens for two companies. His adoration of expensive pens never flagged, even when it took "a royal amount of cussing to make the thing[s] go." Parker 51s they were not.

Still, Twain did more than anyone to ensure the eventual triumph of typewriters over high-end pens. He submitted the first typewritten manuscript to a publisher, *Life on the Mississippi*, in 1883. (It was dictated to a secretary, not typed by Twain.) And when the Remington typewriter company asked him to endorse its machines (Twain had reluctantly bought another), he declined with a crusty letter—which Remington turned around and printed anyway.* Even the acknowledgment that Twain, probably the most popular person in America, owned one was endorsement enough.

These stories of cussing out pens he loved and using

typewriters he hated underline a contradiction in Twain. Though perhaps the antithesis of Goethe in a literary sense, the demotic, democratic Twain shared Goethe's ambivalence about technology. Twain had no pretensions of practicing science, but both he and Goethe were fascinated by scientific discovery. At the same time, they doubted *Homo sapiens* had enough wisdom to use technology properly. In Goethe, this doubt manifested itself as *Faust*. And Twain wrote what we might recognize today as science fiction. Really. In contrast to his laddish riverboat novels, he wrote short stories about inventions, technology, dystopias, space and time travel, even, in his bemusing story "Sold to Satan," the perils of the periodic table.

The story, two thousand words long, starts shortly after a hypothetical crash of steel shares around 1904. The narrator, sick of scrabbling for money, decides to sell his immortal soul to Mephistopheles. To hammer out a deal, he and Satan meet in a dark, unnamed lair at midnight, drink some hot toddies, and discuss the depressingly modest going price for souls. Pretty soon, though, they're sidetracked by an unusual feature of Satan's anatomy—he's made entirely of radium.

Six years before Twain's story, Marie Curie had astounded the scientific world with her tales of radioactive elements. It was genuine news, but Twain must have been pretty plugged into the scientific scene to incorporate all the cheeky details he did into "Sold to Satan." Radium's radioactivity charges the air around it electrically, so Satan glows a luminescent green, to the narrator's delight. Also, like a warm-blooded rock, radium is always hotter than its surroundings, because its radioactivity heats it up. This heat grows exponentially as more radium is concentrated together. As a result, Twain's six-foot-one, "nine-hundred-odd"-pound Satan is hot enough to light a cigar with his fingertip. (He quickly puts it out, though, to "save it for

Voltaire." Hearing this, the narrator makes Satan take fifty more cigars for, among others, Goethe.)

Later, the story goes into some detail about refining radioactive metals. It's far from Twain's sharpest material. But like the best science fiction, it's prescient. To avoid incinerating people he comes across, radium-bodied Satan wears a protective coat of polonium, another new element discovered by Curie. Scientifically, this is rubbish: a "transparent" shell of polonium, "thin as a gelatine film," could never withhold the heat of a critical mass of radium. But we'll forgive Twain, since the polonium serves a larger dramatic purpose. It gives Satan a reason to threaten, "If I should strip off my skin the world would vanish away in a flash of flame and a puff of smoke, and the remnants of the extinguished moon would sift down through space a mere snow-shower of gray ashes!"

Twain being Twain, he could not let the Devil end the story in a position of power. The trapped radium heat is so intense that Satan soon admits, with unintended irony, "I burn. I suffer within." But jokes aside, Twain was already trembling about the awesome power of nuclear energy in 1904. Had he lived forty years more, he surely would have shaken his head—dispirited, yet hardly surprised—to see people lusting after nuclear missiles instead of plentiful atomic energy. Unlike Goethe's forays into hard science, Twain's stories about science can still be read today with instruction.

Twain surveyed the lower realm of the periodic table with despair. But of all the tales of artists and elements, none seems sadder or harsher, or more Faustian, than poet Robert Lowell's adventures with one of the primordial elements, lithium, at the very top of the table.

When they were all youngsters at a prep school in the early

1930s, friends nicknamed Lowell "Cal" after Caliban, the howl-
ing man-beast in *The Tempest*. Others swear Caligula inspired
the epithet. Either way, the name fit the confessional poet, who
exemplified the mad artist—someone like van Gogh or Poe,
whose genius stems from parts of the psyche most of us cannot
access, much less harness for artistic purposes. Unfortunately,
Lowell couldn't rein in his madness outside the margins of his
poems, and his lunacy bled all over his real life. He once turned
up sputtering on a friend's doorstep, convinced that he (Lowell) was the Virgin Mary. Another time, in Bloomington, Indi-
ana, he convinced himself he could stop cars on the highway
by spreading his arms wide like Jesus. In classes he taught, he
wasted hours babbling and rewriting the poems of nonplussed
students in the obsolete style of Tennyson or Milton. When
nineteen, he abandoned a fiancée and drove from Boston to the
country house of a Tennessee poet who Lowell hoped would
mentor him. He just assumed that the man would put him up.
The poet graciously explained there was no room at the inn,
so to speak, and joked that Lowell would have to camp on the
lawn if he wanted to stay. Lowell nodded and left—for Sears.
He bought a pup tent and returned to rough it on the grass.

The literary public delighted in these stories, and during
the 1950s and 1960s, Lowell was the preeminent poet in the
United States, winning prizes and selling thousands of books.
Everyone assumed Lowell's aberrations were the touch of some
madly divine muse. Pharmaceutical psychology, a field coming
into its own in that era, had a different explanation: Cal had
a chemical imbalance, which rendered him manic-depressive.
The public saw only the wild man, not his incapacitating black
moods—moods that left him broken spiritually and increasingly broke financially. Luckily, the first real mood stabi-
lizer, lithium, came to the United States in 1967. A desperate

Lowell—who'd just been incarcerated in a psychiatric ward, where doctors had confiscated his belt and shoelaces—agreed to be medicated.

Curiously, for all its potency as a drug, lithium has no normal biological role. It's not an essential mineral like iron or magnesium, or even a trace nutrient like chromium. In fact, pure lithium is a scarily reactive metal. People's linty pockets have reportedly caught fire when keys or coins short-circuited portable lithium batteries as they jangled down the street. Nor does lithium (which in its drug form is a salt, lithium carbonate) work the way we expect drugs to. We take antibiotics at the height of an infection to knock the microbes out. But taking lithium at the height of mania or in the canyon of depression won't fix the episode. Lithium only prevents the next episode from starting. And although scientists knew about lithium's efficacy back in 1886, until recently they had no clue why it worked.

Lithium tweaks many mood-altering chemicals in the brain, and its effects are complicated. Most interesting, lithium seems to reset the body's circadian rhythm, its inner clock. In normal people, ambient conditions, especially the sun, dictate their humors and determine when they are tuckered out for the day. They're on a twenty-four-hour cycle. Bipolar people run on cycles independent of the sun. And run and run. When they're feeling good, their brains flood them with sunshiny neurostimulants, and a lack of sunshine does not turn the spigot off. Some call it "pathological enthusiasm": such people barely need sleep, and their self-confidence swells to the point that a Bostonian male in the twentieth century can believe that the Holy Spirit has chosen him as the vessel of Jesus Christ. Eventually, those surges deplete the brain, and people crash. Severe manic-depressives, when the "black dogs" have them, sometimes take to bed for weeks.

Lithium regulates the proteins that control the body's inner clock. This clock runs, oddly, on DNA, inside special neurons deep in the brain. Special proteins attach to people's DNA each morning, and after a fixed time they degrade and fall off. Sunlight resets the proteins over and over, so they hold on much longer. In fact, the proteins fall off only after darkness falls—at which point the brain should "notice" the bare DNA and stop producing stimulants. This process goes awry in manic-depressives because the proteins, despite the lack of sunlight, remain bound fast to their DNA. Their brains don't realize they should stop revving. Lithium helps cleave the proteins from DNA so people can wind down. Notice that sunlight still trumps lithium during the day and resets the proteins; it's only when the sunlight goes away at night that lithium helps DNA shake free. Far from being sunshine in a pill, then, lithium acts as "anti-sunlight." Neurologically, it undoes sunlight and thereby compresses the circadian clock back to twenty-four hours—preventing both the mania bubble from forming and the Black Tuesday crash into depression.

Lowell responded immediately to lithium. His personal life grew steadier (though by no means steady), and at one point he pronounced himself cured. From his new, stable perspective, he could see how his old life—full of fights, drinking binges, and divorces—had laid waste to so many people. For all his frank and moving lines within his poems, nothing Lowell ever wrote was as poignant—and nothing about the fragile chemistry of human beings was as moving—as a simple plaint to his publisher, Robert Giroux, after doctors started him on lithium.

"It's terrible, Bob," he said, "to think that all I've suffered, and all the suffering I've caused, might have arisen from the lack of a little salt in my brain."

Lowell felt that his life improved on lithium, yet the effect of lithium on his art was debatable. As Lowell did, most artists feel that trading a manic-depressive cycle for a muted, prosaic circadian rhythm allows them to work productively without being distracted by mania or sedated by depression. There's always been debate, though, about whether their work suffers after their "cure," after they've lost access to that part of the psyche most of us never glimpse.

Many artists report feeling flatlined or tranquilized on lithium. One of Lowell's friends reported that he looked like something transported around in a zoo. And his poetry undoubtedly changed after 1967, growing rougher and purposely less polished. He also, instead of inventing lines from his wild mind, began poaching lines from private letters, which outraged the people he quoted. Such work won Lowell a Pulitzer Prize in 1974, but it hasn't weathered well. Especially compared with his vivacious younger work, it's barely read today. For all that the periodic table inspired Goethe, Twain, and others, Lowell's lithium may be a case where it provided health but subdued art, and made a mad genius merely human.

15

An Element of Madness

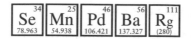

Robert Lowell typified the mad artist, but there's another psychological deviant in our collective cultural psyche: the mad scientist. The mad scientists of the periodic table tended to have fewer public outbursts than mad artists, and they generally didn't lead notorious private lives either. Their psychological lapses were subtler, and their mistakes were typical of a peculiar kind of madness known as pathological science.* And what's fascinating is how that pathology, that madness, could exist side by side in the same mind with brilliance.

Unlike virtually every other scientist in this book, William Crookes, born to a tailor in London in 1832, never worked at a university. The first of sixteen children, he later fathered ten of his own, and he supported his enormous family by writing a popular book on diamonds and editing a bumptious, gossipy journal of science goings-on, *Chemical News*. Nevertheless, Crookes—a bespectacled man with a beard and pointy mustache—did enough world-class science on elements such as selenium and thallium to get elected to England's premier scientific club, the Royal Society, at just thirty-one years of age. A decade later, he was almost kicked out.

His fall began in 1867, when his brother Philip died at sea.*

Despite, or perhaps because of, their abundance of family, William and the other Crookeses nearly went mad with grief. At the time, spiritualism, a movement imported from America, had overrun the houses of aristocrats and shopkeeps alike all over England. Even someone like Sir Arthur Conan Doyle, who invented the hyperrationalist detective Sherlock Holmes, could find room in his capacious mind to accept spiritualism as genuine. Products of their time, the Crookes clan—mostly tradesmen with neither scientific training nor instinct—began attending séances en masse to comfort themselves and to chat with poor departed Philip.

It's not clear why William tagged along one night. Perhaps solidarity. Perhaps because another brother was stage manager for the medium. Perhaps to dissuade everyone from going back— privately, in his diary, he had dismissed such spiritual "contact" as fraudulent pageantry. Yet watching the medium play the accordion with no hands and write "automatic messages," Ouija board–style, with a stylus and plank impressed the skeptic despite himself. His defenses were lowered, and when the medium began relaying babbled messages from Philip in the great beyond, William began bawling. He went to more sessions, and even invented a scientific device to monitor the susurrus of wandering spirits in the candlelit rooms. It's not clear if his new radiometer—an evacuated glass bulb with a very sensitive weather vane inside—actually detected Philip. (We can hazard a guess.) But William couldn't dismiss what he felt holding hands with family members at the meetings. His attendance became regular.

Such sympathies put Crookes in the minority among his fellow rationalists in the Royal Society—probably a minority of one. Mindful of this, Crookes concealed his biases in 1870 when he announced that he had drawn up a scientific study of spiritualism, and most fellows of the Royal Society were delighted, assuming he would demolish the whole scene in his rowdy journal. Things did not turn out so neatly. After three years of chanting and summoning, Crookes published "Notes of an Enquiry into the Phenomena Called Spiritual" in 1874 in a journal he owned called the *Quarterly Journal of Science*. He compared himself to a traveler in exotic lands, a Marco Polo of the paranormal. But instead of attacking all the spiritual-ist mischief—"levitation," "phantoms," "percussive sounds," "luminous appearances," "the rising of tables and chairs off the ground"—he concluded that neither charlatanism nor mass hypnosis could explain (or at least not *wholly* explain) all he'd seen. It wasn't an uncritical endorsement, but Crookes did claim to find a "residual" of legitimate supernatural forces.*

Coming from Crookes, even such tepid support shocked everyone in England, including spiritualists. Recovering quickly, they began shouting hosannas about Crookes from the mountaintops. Even today, a few ghost hunters haul out his creaky paper as "proof" that smart people will come around to spiritualism if they approach it with an open mind. Crookes's fellows in the Royal Society were equally surprised but rather more aghast. They argued that Crookes had been blinded by parlor tricks, swept up in crowd dynamics, and charmed by charismatic gurus. They also tore into the dubious scientific veneer he'd given his report. Crookes had recorded irrelevant "data" about the temperature and barometric pressure inside the medium's lair, for instance, as if immaterial beings wouldn't poke their heads out in inclement weather. More uncomfort-ably, former friends attacked Crookes's character, calling him a rube, a shill. If spiritualists sometimes cite Crookes today, a few scientists still cannot forgive him for enabling 135 years of New Age-y BS. They even cite his work on the elements as proof he went crazy.

When young, you see, Crookes had pioneered the study of selenium. Though an essential trace nutrient in all animals (in humans, the depletion of selenium in the bloodstream of AIDS patients is a fatally accurate harbinger of death), selenium is toxic in large doses. Ranchers know this well. If not watched carefully, their cattle will root out a prairie plant of the pea family known as locoweed, some varieties of which sponge up selenium from the soil. Cattle that munch on locoweed begin to stagger and stumble and develop fevers, sores, and anorexia—a suite of symptoms known as the blind staggers. Yet they enjoy the high. In the surest sign that selenium actually makes them go mad, cattle grow addicted to locoweed despite its awful side effects and eat it to the exclusion of anything else. It's animal meth. Some imaginative historians even pin Custer's loss at the Battle of the Little Bighorn on his horses' taking hits of loco before the battle. Overall, it's fitting that "selenium" comes from *selene*, Greek for "moon," which has links—through *luna*, Latin for "moon"—to "lunatic" and "lunacy."

Given that toxicity, it might make sense to retroactively blame Crookes's delusions on selenium. Some inconvenient facts undermine that diagnosis, though. Selenium often attacks within a week; Crookes got goofy in early middle age, long after he'd stopped working with selenium. Plus, after decades of ranchers' cursing out element thirty-four every time a cow stumbled, many biochemists now think that other chemicals in locoweed contribute just as much to the craziness and intoxication. Finally, in a clinching clue, Crookes's beard never fell out, a classic symptom of selenosis.

A full beard also argues against his being driven mad, as some have suggested, by another depilatory on the periodic table—the poisoner's poison, thallium. Crookes discovered thallium at age twenty-six (a finding that almost ensured his election to the Royal Society) and continued to play with it in his lab for a decade. But he apparently never inhaled enough even to lose his whiskers. Besides, would someone ravished by thallium (or selenium) retain such a sharp mind into old age? Crookes actually withdrew from spiritualist circles after 1874, rededicating himself to science, and major discoveries lay ahead. He was the first to suggest the existence of isotopes. He built vital new scientific equipment and confirmed the presence of helium in rocks, its first detection on earth. In 1897, the newly knighted Sir William dove into radioactivity, even discovering (though without realizing it) the element protactinium in 1900.

No, the best explanation for Crookes's lapse into spiritualism is psychological: ruined by grief for his brother, he succumbed, avant la lettre, to pathological science.

In explaining what pathological science is, it's best to clear up any misconceptions about that loaded word, "pathological," and explain up front what pathological science is *not*. It's not fraud, since the adherents of a pathological science believe they're right—if only everyone else could see it. It's not pseudoscience, like Freudianism and Marxism, fields that poach on the imprimatur of science yet shun the rigors of the scientific method. It's also not politicized science, like Lysenkoism, where people swear allegiance to a false science because of threats or a skewed ideology. Finally, it's not general clinical madness or merely deranged belief. It's a particular madness, a meticulous and scientifically informed delusion. Pathological scientists pick out a marginal and unlikely phenomenon that appeals to them for whatever reason and bring all their scientific acumen to proving its existence. But the game is rigged from the start: their science serves only the deeper emotional need to believe in something. Spiritualism per se isn't a pathological science, but it became so in Crookes's hands because of his careful "experiments" and the scientific trimmings he gave the experiments.

And actually, pathological science doesn't always spring from fringe fields. It also thrives in legitimate but speculative fields, where data and evidence are scarce and hard to interpret. For example, the branch of paleontology concerned with reconstructing dinosaurs and other extinct creatures provides another great case study in pathological science.

At some level, of course, we don't know squat about extinct creatures: a whole skeleton is a rare find, and soft tissue impressions are vanishingly rare. A joke among people who reconstruct paleofauna is that if elephants had gone extinct way back when, anyone who dug up a mammoth skeleton today would conjure up a giant hamster with tusks, not a woolly pachyderm with a trunk. We'd know just as little about the glories of other animals as well—stripes, waddles, lips, potbellies, belly buttons, snouts, gizzards, four-chambered stomachs, and humps, not to mention their eyebrows, buttocks, toenails, cheeks, tongues, and nipples. Nevertheless, by comparing the grooves and depressions on fossilized bones with modern creatures' bones, a trained eye can figure out the musculature, enervation, size, gait, dentition, and even mating habits of extinct species. Paleontologists just have to be careful about extrapolating too far.

A pathological science takes advantage of that caution. Basically, its believers use the ambiguity about evidence *as* evidence—claiming that scientists don't know everything and therefore there's room for my pet theory, too. That's exactly what happened with manganese and the megalodon.*

This story starts in 1873, when the research vessel HMS *Challenger* set out from England to explore the Pacific Ocean. In a wonderfully low-tech setup, the crew dropped overboard huge buckets tied to ropes three miles long and dredged the ocean floor. In addition to fantastical fish and other critters, they hauled up dozens upon dozens of spherical rocks shaped like fossilized potatoes and also fat, solid, mineralized ice cream cones. These hunks, mostly manganese, appeared all over the seabed in every part of the ocean, meaning there had to be untold billions of them scattered around the world.

That was the first surprise. The second took place when the crew cracked open the cones: the manganese had formed itself around giant shark teeth. The biggest, most pituitarily freakish shark teeth today run about two and a half inches max. The manganese-covered teeth stretched five or more inches—mouth talons capable of shattering bone like an ax. Using the same basic techniques as with dinosaur fossils, paleontologists determined (just from the teeth!) that this Jaws[3], dubbed the megalodon, grew to approximately fifty feet, weighed approximately fifty tons, and could swim approximately fifty miles per hour. It could probably close its mouth of 250 teeth with a megaton force, and it fed mostly on primitive whales in shallow, tropical waters. It probably died out as its prey migrated permanently to colder, deeper waters, an environment that didn't suit its high metabolism and ravenous appetite.

All fine science so far. The pathology started with the manganese.* Shark teeth litter the ocean floor because they're about the hardest biological substance known, the only part of shark carcasses that survive the crush of the deep ocean (most sharks have cartilaginous skeletons). It's not clear why manganese, of all the dissolved metals in the ocean, galvanizes shark teeth, but scientists know roughly how quickly it accumulates: between one-half and one and a half millimeters per millennium. From that rate they have determined that the vast majority of recovered teeth date from at least 1.5 million years ago, meaning the megalodons probably died out around then.

But—and here was the gap into which some people rushed—some megalodon teeth had mysteriously thin manganese plaque, about eleven thousand years' worth. Evolutionarily, that's an awfully short time. And really, what's to say scientists won't soon find one from ten thousand years ago? Or eight thousand years ago? Or later?

You can see where this thinking leads. In the 1960s, a few enthusiasts with *Jurassic Park* imaginations grew convinced that rogue megalodons still lurk in the oceans. "Megalodon lives!" they cried. And like rumors about Area 51 or the Kennedy assassination, the legend has never quite died. The most common tale is that megalodons have evolved to become deep-sea divers and now spend their days fighting krakens in the black depths. Reminiscent of Crookes's phantoms, megalodons are *supposed* to be elusive, which gives people a convenient escape when pressed on why the giant sharks are so scarce nowadays.

There's probably not a person alive who, deep down, doesn't hope that megalodons still haunt the seas. Unfortunately, the idea crumbles under scrutiny. Among other things, the teeth with thin layers of manganese were almost certainly torn up from old bedrock beneath the ocean floor (where they accumulated no manganese) and exposed to water only recently. They're probably much older than eleven thousand years. And although there have been eyewitness accounts of the beasts, they're all from sailors, notorious storytellers, and the megalodons in their stories vary manically in size and shape. One all-white Moby Dick shark stretched up to three hundred feet long! (Funny, though, no one thought to snap a picture.) Overall, such stories, as with Crookes's testimony about supernatural beings, depend on subjective interpretations, and without objective evidence, it's not plausible to conclude that megalodons, even a few of them, slipped through evolution's snares.

But what really makes the ongoing hunt for megalodons pathological is that doubt from the establishment only deepens people's convictions. Instead of refuting the manganese findings, they counterattack with heroic tales of rebels, rogues who proved squaresville scientists wrong in the past. They invariably bring up the coelacanth, a primitive deep-sea fish once thought to have gone extinct eighty million years ago, until it turned up in a fish market in South Africa in 1938. According to this logic, because scientists were wrong about the coelacanth, they might be wrong about the megalodon, too. And "might" is all the megalodon lovers need. For their theories about its survival aren't based on a preponderance of evidence, but on an emotional attachment: the hope, the need, for something fantastic to be true.

There's probably no better example of such emotion than in the next case study—that all-time-great pathological science, that Alamo for true believers, that seductress of futurists, that scientific hydra: cold fusion.

Pons and Fleischmann. Fleischmann and Pons. It was supposed to be the greatest scientific duo since Watson and Crick, perhaps stretching back to Marie and Pierre Curie. Instead, their fame rotted into infamy. Now the names B. Stanley Pons and Martin Fleischmann evoke only, however unfairly, thoughts of impostors, swindlers, and cheats.

The experiment that made and unmade Pons and Fleischmann was, so to speak, deceptively simple. The two chemists, headquartered at the University of Utah in 1989, placed a palladium electrode in a chamber of heavy water and turned on a current. Running a current through regular water will shock the H_2O and produce hydrogen and oxygen gas. Something similar happened in the heavy water, except the hydrogen in heavy water has an extra neutron. So instead of normal hydrogen gas (H_2) with two protons total, Pons and Fleischmann created molecules of hydrogen gas with two protons and two neutrons.

What made the experiment special was the combination of heavy hydrogen with palladium, a whitish metal with one flabbergasting property: it can swallow nine hundred times its own volume of hydrogen gas. That's roughly equivalent to a 250-pound man swallowing a dozen African bull elephants* and not gaining an inch on his waistline. And as the palladium electrode in the heavy water started to pack in hydrogen, Pons and Fleischmann's thermometers and other instruments spiked. The water got far warmer than it should have, than it *could* have, given the meager energy of the incoming current. Pons reported that during one really good spike, his superheated H_2O burned a hole in a beaker, the lab bench beneath it, and the concrete floor beneath that.

Or at least they got spikes sometimes. Overall, the experiment was erratic, and the same setup and trial runs didn't always produce the same results. But rather than nail down what was happening with the palladium, the two men let their fancies convince them they had discovered cold fusion—fusion that didn't require the incredible temperatures and pressures of stars, but took place at room temperature. Because palladium could cram so much heavy hydrogen inside it, they guessed it somehow fused its protons and neutrons into helium, releasing gobs of energy in the process.

Rather imprudently, Pons and Fleischmann called a press conference to announce their results, basically implying that the world's energy problems were over, cheaply and without pollution. And somewhat like palladium itself, the media swallowed the grandiose claim. (It soon came out that another Utahan, physicist Steven Jones, had pursued similar fusion experiments. Jones fell into the background, however, since he made more modest claims.) Pons and Fleischmann became instant celebrities, and the momentum of public opinion appeared to sway even scientists. At an American Chemical Society meeting shortly after the announcement, the duo received a standing ovation.

But there's some important context here. In applauding Fleischmann and Pons, many scientists were probably really thinking about superconductors. Until 1986, superconductors were thought to be flat-out impossible above −400°F. Suddenly, two German researchers—who would win the Nobel Prize in record time, a year later—discovered superconductors that worked above that temperature. Other teams jumped in and within a few months had discovered "high-temperature" yttrium superconductors that worked at −280°F. (The record today stands at −218°F.) The point is that many scientists who'd predicted the impossibility of such superconductors felt like asses. It was the physics equivalent of finding the coelacanth. And like megalodon romantics, cold-fusion lovers in 1989 could point to the recent superconductor craziness and force normally dismissive scientists to suspend judgment. Indeed, cold-fusion fanatics seemed giddy at the chance to overthrow old dogma, a delirium typical of pathological science.

Still, a few skeptics, especially at Cal Tech, seethed. Cold fusion upset these men's scientific sensibilities, and Pons and Fleischmann's arrogance upset their modesty. The two had bypassed the normal peer-review process in announcing results, and some considered them charlatans intent on enriching themselves, especially after they appealed directly to President George H. W. Bush for $25 million in immediate research funds. Pons and Fleischmann didn't help matters by refusing to

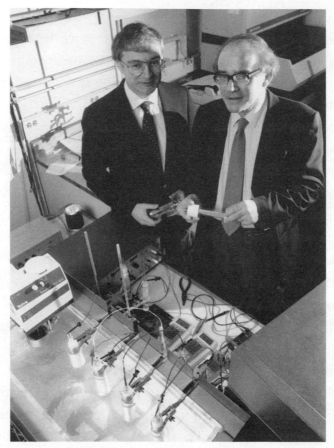

Despite withering dismissals from nearly every other scientist on earth, Stanley Pons and Martin Fleischmann claimed they had produced cold fusion at room temperature. Their apparatus consisted of a heavy-water bath with electrodes made of the superabsorbent element palladium. (Special Collections Department, J. Willard Marriott Library, University of Utah)

answer—as if such inquiries were insulting—questions about their palladium apparatus and experimental protocol. They claimed they didn't want their ideas to be stolen, but it sure looked as if they were hiding something.

Nevertheless, increasingly doubtful scientists across the world (except in Italy, where yet another cold-fusion claim popped up) learned enough from what the two men said to rig up their own palladium and heavy-hydrogen experiments, and they began pummeling the Utah scientists with null results. A few weeks later, after perhaps the most concerted effort since Galileo to discredit, even disgrace, scientists, hundreds of chemists and physicists held what amounted to an anti–Pons and Fleischmann rally in Baltimore. They showed, embarrassingly, that the duo had overlooked experimental errors and used faulty measuring techniques. One scientist suggested that the two had let the hydrogen gas build up and that their biggest "fusion" spikes were chemical explosions, à la the *Hindenburg*. (The supposed fusion spike that burned holes in the table and bench happened overnight, when no one was around.) Usually it takes years to root out a scientific error, or at least to resolve a controversial question, but cold fusion was cold and dead within forty days of the initial announcement. One wag who attended the conference summed up the brouhaha in biting, if unrhythmical, verse:

Tens of millions of dollars at stake, Dear Brother
Because some scientists put a thermometer
At one place and not another.

But the psychologically interesting parts of the affair were still to come. The need to believe in clean, cheap energy for the whole world proved tenacious, and people could not still their heartstrings so quickly. At this point, the science mutated into something pathological. As with investigations into the paranormal, only a guru (the medium, or Fleischmann and Pons) had the power to produce the key results, and only under contrived circumstances, never in the open. That didn't give pause to and in fact only encouraged cold-fusion enthusiasts. For their part, Pons and Fleischmann never backed down, and their followers defended the two (not to mention themselves) as important rebels, the only people who *got it*. Some critics countered with their own experiments for a while after 1989, but cold fusionists always explained away any damning results, sometimes with more ingenuity than they showed in their original scientific work. So the critics eventually gave up. David Goodstein, a Cal Tech physicist, summed things up in an excellent essay on cold fusion: "Because the Cold-Fusioners see themselves as a community under siege, there is little internal criticism. Experiments and theories tend to be accepted at face value, for fear of providing even more fuel for external critics, if anyone outside the group was bothering to listen. In these circumstances, crackpots flourish, making matters worse for those who believe that there is serious science going on here." It's hard to imagine a better, more concise description of pathological science.*

The most charitable explanation of what happened to Pons and Fleischmann is this. It seems unlikely they were charlatans who knew that cold fusion was bunkum but wanted a quick score. It wasn't 1789, where they could have just skedaddled and scammed the rubes in the next town over. They were going to get caught. Maybe they had doubts but were blinded by ambition and wanted to see what it felt like to be brilliant in the world's eyes, even for a moment. Probably, though, these two men were just misled by a queer property of palladium. Even today, no one knows how palladium guzzles so much hydrogen. In a *slight* rehabilitation of Pons and Fleischmann's work (though not their interpretation of it), some scientists do think that something funny is going on in palladium–heavy water experiments. Strange bubbles appear in the metal, and its atoms rearrange themselves in novel ways. Perhaps even some weak nuclear forces are involved. To their credit, Pons and Fleischmann pioneered this work. It's just not what they wanted to, or will, go down in science history for.

Not every scientist with a touch of madness ends up drowning in pathological science, of course. Some, like Crookes, escape and go on to do great work. And then there are the rare cases where what seems like pathological science at the outset turns out to be legitimate. Wilhelm Röntgen tried his damnedest to prove himself wrong while pursuing a radical discovery about invisible rays, but couldn't. And because of his persistence and insistence on the scientific method, this mentally fragile scientist really did rewrite history.

In November 1895, Röntgen was playing around in his laboratory in central Germany with a Crookes tube, an important new tool for studying subatomic phenomena. Named after its inventor, you know who, the Crookes tube consisted of an evacuated glass bulb with two metal plates inside at either end. Running a current between the plates caused a beam to leap across the vacuum, a crackle of light like something from a special effects lab. Scientists now know it's a beam of electrons, but in 1895 Röntgen and others were trying to figure that out.

A colleague of Röntgen's had found that when he made a Crookes tube with a small aluminium foil window (reminiscent of the titanium window Per-Ingvar Brånemark later welded onto rabbit bones), the beam would tunnel through the foil into the air. It died pretty quickly—air was like poison to the beam—but it could light up a phosphorescent screen a few inches distant. A little neurotically, Röntgen insisted on repeating all his colleagues' experiments no matter how minor, so he built this setup himself in 1895, but with some alterations. Instead of leaving his Crookes tube naked, he covered it with black paper, so that the beam would escape only through the foil. And instead of the phosphorescing chemicals his colleague had used, he painted his plates with a luminescent barium compound.

Accounts of what happened next vary. As Röntgen was running some tests, making sure his beam jumped between the plates properly, something caught his attention. Most accounts say it was a piece of cardboard coated with barium, which he'd propped on a nearby table. Other contemporary accounts say it was a piece of paper that a student had finger-painted with barium, playfully drawing the letter *A* or *S*. Regardless, Röntgen, who was color-blind, would have seen just a dance of white on the edge of his vision at first. But every time he turned the current on, the barium plate (or the letter) glowed.

Röntgen confirmed that no light was escaping from the blackened Crookes tube. He'd been sitting in a dark lab, so sunshine couldn't have caused the sparkle either. But he also knew the Crookes beams couldn't survive long enough in air to jump over to the plate or letter. He later admitted he thought he was hallucinating—the tube was clearly the cause, but he knew of nothing that could warp through opaque black paper.

So he propped up a barium-coated screen and put the nearest objects at hand, like a book, near the tube to block the beam. To his horrified amazement, an outline of a key he used as a bookmark appeared on the screen. He could somehow *see through things*. He tried objects in closed wooden boxes and saw through those, too. But the truly creepy, truly black-magic moment came when he held up a plug of metal—and saw the bones of his own hand. At this point, Röntgen ruled out mere hallucination. He assumed he'd gone stark mad.

We can laugh today at his getting so worked up over discovering X-rays. But notice his remarkable attitude here. Instead of leaping to the convenient conclusion that he'd discovered something radically new, Röntgen assumed he'd made a mistake somewhere. Embarrassed, and determined to prove himself wrong, he locked himself in his lab, isolating himself for seven unrelenting weeks in his cave. He dismissed his assistants and took his meals grudgingly, gulping down food and grunting more than talking to his family. Unlike Crookes, or the megalodon hunters, or Pons and Fleischmann, Röntgen labored heroically to fit his findings in with known physics. He didn't want to be revolutionary.

Ironically, though he did everything to skirt pathological science, Röntgen's papers show that he couldn't shake the thought he had gone mad. Moreover, his muttering and his uncharacteristic temper made other people question his sanity. He jokingly said to his wife, Bertha, "I'm doing work that will make people say, 'Old Röntgen has gone crazy!'" He was fifty then, and she must have wondered.

Still, the Crookes tube lit up the barium plates every time, no matter how much he disbelieved. So Röntgen began documenting the phenomenon. Again, unlike the three pathological cases above, he dismissed any fleeting or erratic effects, anything that might be considered subjective. He sought only objective results, like developed photographic plates. At last, slightly more confident, he brought Bertha into the lab one afternoon and exposed her hand to the X-rays. Upon seeing her bones, she freaked out, thinking it a premonition of her death. She refused to go back into his haunted lab after that, but her reaction brought immeasurable relief to Röntgen. Possibly the most loving thing Bertha ever did for him, it proved he hadn't imagined everything.

At that point, Röntgen emerged, haggard, from his laboratory and informed his colleagues across Europe about "röntgen rays." Naturally, they doubted him, just as they'd scorned Crookes and later scientists would scorn the megalodon and

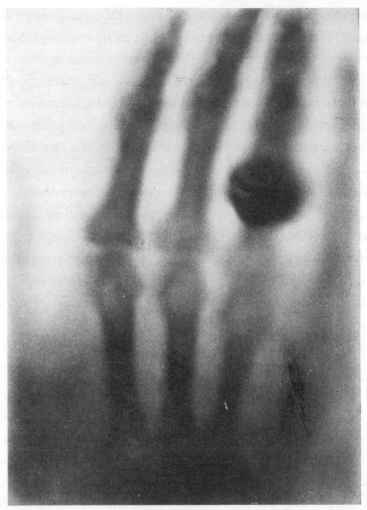

This early X-ray revealed the bones and impressive ring of Bertha Röntgen, wife of Wilhelm Röntgen. Wilhelm, who feared he'd gone mad, was relieved when his wife also saw the bones of her hand on a barium-coated plate. She, less sanguine, thought it an omen of death.

cold fusion. But Röntgen had been patient and modest, and every time someone objected, he countered by saying he'd already investigated that possibility, until his colleagues had no more objections. And herein lies the uplifting side to the normally severe tales of pathological science.

Scientists can be cruel to new ideas. One can imagine them asking, "What sort of 'mystery beams' can fly invisibly through black paper, Wilhelm, and light up the bones in your body? *Bah*." But when he fought back with solid proof, with repeatable experiments, most overthrew their old ideas to embrace his. Though a middling professor his whole life, Röntgen became every scientist's hero. In 1901, he won the inaugural Nobel Prize in Physics. Two decades later, a physicist named Henry Moseley used the same basic X-ray setup to revolutionize the study of the periodic table. And people were still so smitten a century later that in 2004, the largest official element on the periodic table at the time, number 111, long called unununium, became roentgenium.

Part V

ELEMENT SCIENCE TODAY AND TOMORROW

16

Chemistry Way, Way Below Zero

Röntgen not only provided an example of brilliantly meticulous science; he also reminded scientists that the periodic table is never empty of surprises. There's always something novel to discover about the elements, even today. But with most of the easy pickings already plucked by Röntgen's time, making new discoveries required drastic measures. Scientists had to interrogate the elements under increasingly severe conditions—especially extreme cold, which hypnotizes them into strange behaviors. Extreme cold docsn't always portend well for the humans making the discoveries either. While the latter-day heirs of Lewis and Clark had explored much of Antarctica by 1911, no human being had ever reached the South Pole. Inevitably, this led to an epic race among explorers to get there first—which led just as inevitably to a grim cautionary tale about what can go wrong with chemistry at extreme temperatures.

That year was chilly even by Antarctic standards, but a band of pale Englishmen led by Robert Falcon Scott nonetheless determined that they would be the first to reach ninety degrees south latitude. They organized their dogs and supplies, and a caravan set off in November. Much of the caravan was a support team, which cleverly dropped caches of food and fuel on the way out so that the small final team that would dash to the pole could retrieve them on the way back.

Little by little, more of the caravan peeled off, and finally, after slogging along for months on foot, five men, led by Scott, arrived at the pole in January 1912—only to find a brown pup tent, a Norwegian flag, and an annoyingly friendly letter. Scott had lost out to Roald Amundsen, whose team had arrived a month earlier. Scott recorded the moment curtly in his diary: "The worst has happened. All the daydreams must go." And shortly afterward: "Great God! This is an awful place. Now for the run home and a desperate struggle. I wonder if we can do it."

Dejected as Scott's men were, their return trip would have been difficult anyway, but Antarctica threw up everything it could to punish and harass them. They were marooned for weeks in a monsoon of snow flurries, and their journals (discovered later) showed that they faced starvation, scurvy, dehydration, hypothermia, and gangrene. Most devastating was the lack of heating fuel. Scott had trekked through the Arctic the year before and had found that the leather seals on his canisters of kerosene leaked badly. He'd routinely lost half of his fuel. For the South Pole run, his team had experimented with tin-enriched and pure tin solders. But when his bedraggled men reached the canisters awaiting them on the return trip, they found many of them empty. In a double blow, the fuel had often leaked onto foodstuffs.

Without kerosene, the men couldn't cook food or melt ice to drink. One of them took ill and died; another went insane in the cold and wandered off. The last three, including Scott, pushed on. They officially died of exposure in late March 1912, eleven miles wide of the British base, unable to get through the last nights.

In his day, Scott had been as popular as Neil Armstrong—Britons received news of his plight with gnashing of teeth, and one church even installed stained-glass windows in his honor in 1915. As a result, people have always sought an excuse to absolve him of blame, and the periodic table provided a convenient villain. Tin, which Scott used as solder, has been a prized metal since biblical times because it's so easy to shape. Ironically, the better metallurgists got at refining tin and purifying it, the worse it became for everyday use. Whenever pure tin tools or tin coins or tin toys got cold, a whitish rust began to creep over them like hoarfrost on a window in winter. The white rust would break out into pustules, then weaken and corrode the tin, until it crumbled and eroded away.

Unlike iron rust, this was not a chemical reaction. As scientists now know, this happens because tin atoms can arrange themselves inside a solid in two different ways, and when they get cold, they shift from their strong "beta" form to the crumbly, powdery "alpha" form. To visualize the difference, imagine stacking atoms in a huge crate like oranges. The bottom of the crate is lined with a single layer of spheres touching only tangentially. To fill the second, third, and fourth layers, you might balance each atom right on top of one in the first layer. That's one form, or crystal structure. Or you might nestle the second layer of atoms into the spaces between the atoms in the first layer, then the third layer into the spaces between the atoms in the second layer, and so on. That makes a second crystal structure with a different density and different properties. These are just two of the many ways to pack atoms together.

What Scott's men (perhaps) found out the hard way is that an element's atoms can spontaneously shift from a weak crystal to a strong one, or vice versa. Usually it takes extreme conditions to promote rearrangement, like the subterranean heat and pressure that turn carbon from graphite into diamonds. Tin becomes protean at 56°F. Even a sweater evening in October can start the pustules rising and the hoarfrost creeping, and colder temperatures accelerate the process. Any abusive treatment or deformation (such as dents from canisters being tossed onto hard-packed ice) can catalyze the reaction, too, even in tin that is otherwise immune. Nor is this merely a topical defect, a surface scar. The condition is sometimes called tin leprosy because it burrows deep inside like a disease. The alpha–beta shift can even release enough energy to cause audible groaning—vividly called tin scream, although it sounds more like stereo static.

The alpha–beta shift of tin has been a convenient chemical scapegoat throughout history. Various European cities with harsh winters (e.g., St. Petersburg) have legends about expensive tin pipes on new church organs exploding into ash the instant the organist blasted his first chord. (Some pious citizens were more apt to blame the Devil.) Of more world historical consequence, when Napoleon stupidly attacked Russia during the winter of 1812, the tin clasps on his men's jackets reportedly (many historians dispute this) cracked apart and left the Frenchmen's inner garments exposed every time the wind kicked up. As with the horrible circumstances faced by Scott's little band, the French army faced long odds in Russia anyway. But element fifty's changeling ways perhaps made things tougher, and impartial chemistry proved an easier thing to blame* than a hero's bad judgment.

There's no doubt Scott's men found empty canisters— that's in his diary—but whether the disintegration of the tin solder caused the leaks is disputed. Tin leprosy makes so much sense, yet canisters from other teams discovered decades later retained their solder seals. Scott did use purer tin—although it would have to have been extremely pure for leprosy to take hold. Yet no other good explanation besides sabotage exists, and there's no evidence of foul play. Regardless, Scott's little band perished on the ice, victims at least in part of the periodic table.

Quirky things happen when matter gets very cold and shifts from one state to another. Schoolchildren learn about just three interchangeable states of matter—solid, liquid, and gas. High school teachers often toss in a fourth state, plasma, a superheated condition in stars in which electrons detach from their nucleic moorings and go roaming.* In college, students get exposed to superconductors and superfluid helium. In graduate school, professors sometimes challenge students with states such as quark-gluon plasma or degenerate matter. And along the way, a few wiseacres always ask why Jell-O doesn't count as its own special state. (The answer? Colloids like Jell-O are blends of two states.* The water and gelatin mixture can either be thought of as a highly flexible solid or a very sluggish liquid.)

The point is that the universe can accommodate far more states of matter—different micro-arrangements of particles—than are dreamed of in our provincial categories of solid, liquid, and gas. And these new states aren't hybrids like Jell-O. In some cases, the very distinction between mass and energy breaks down. Albert Einstein uncovered one such state while fiddling around with a few quantum mechanics equations in 1924—then dismissed his calculations and disavowed his theoretical discovery as too bizarre to ever exist. It remained impossible, in fact, until someone made it in 1995.

In some ways, solids are the most basic state of matter. (To be scrupulous, the vast majority of every atom sits empty, but the ultra-quick hurry of electrons gives atoms, to our dull senses, the persistent illusion of solidity.) In solids, atoms line up in a repetitive, three-dimensional array, though even the most blasé solids can usually form more than one type of crystal. Scientists can now coax ice into forming fifteen distinctly shaped crystals by using high-pressure chambers. Some ices sink rather than float in water, and others form not six-sided snowflakes, but shapes like palm leaves or heads of cauliflower. One alien ice, Ice X, doesn't melt until it reaches 3,700°F. Even chemicals as impure and complicated as chocolate form quasi-crystals that can shift shapes. Ever opened an old Hershey's Kiss and found it an unappetizing tan? We might call that chocolate leprosy, caused by the same alpha–beta shifts that doomed Scott in Antarctica.

Crystalline solids form most readily at low temperatures, and depending on how low the temperature gets, elements you thought you knew can become almost unrecognizable. Even the aloof noble gases, when forced into solid form, decide that huddling together with other elements isn't such a bad idea. Violating decades of dogma, Canadian-based chemist Neil Bartlett created the first noble gas compound, a solid orange crystal, with xenon in 1962.* Admittedly, this took place at room temperature, but only with platinum hexafluoride, a chemical about as caustic as a superacid. Plus xenon, the largest stable inert gas, reacts far more easily than the others because its electrons are only loosely bound to its nucleus. To get smaller, closed-rank noble gases to react, chemists had to drastically screw down the temperature and basically anesthetize them. Krypton put up a good fight until about –240°F, at which point super-reactive fluorine can latch onto it.

Getting krypton to react, though, was like mixing baking soda and vinegar compared with the struggle to graft something onto argon. After Bartlett's xenon solid in 1962 and the first krypton solid in 1963, it took thirty-seven frustrating years until Finnish scientists finally pieced together the right procedure for argon in 2000. It was an experiment of Fabergé delicacy, requiring solid argon; hydrogen gas; fluorine gas; a highly reactive starter compound, cesium iodide, to get the reaction going; and well-timed bursts of ultraviolet light, all set to bake at a frigid −445°F. When things got a little warmer, the argon compound collapsed.

Nevertheless, below that temperature argon fluorohydride was a durable crystal. The Finnish scientists announced the feat in a paper with a refreshingly accessible title for a scientific work, "A Stable Argon Compound." Simply announcing what they'd done was bragging enough. Scientists are confident that even in the coldest regions of space, tiny helium and neon have never bonded with another element. So for now, argon wears the title belt for the single hardest element humans have forced into a compound.

Given argon's reluctance to change its habits, forming an argon compound was a major feat. Still, scientists don't consider noble gas compounds, or even alpha–beta shifts in tin, truly different states of matter. Different states require appreciably different energies, in which atoms interact in appreciably different ways. That's why solids, where atoms are (mostly) fixed in place; liquids, where particles can flow around each other; and gases, where particles have the freedom to carom about, *are* distinct states of matter.

Still, solids, liquids, and gases have lots in common. For one, their particles are well-defined and discrete. But that sovereignty gives way to anarchy when you heat things up to the plasma state and atoms start to disintegrate, or when you cool things down enough and collectivist states of matter emerge, where the particles begin to overlap and combine in fascinating ways.

Take superconductors. Electricity consists of an easy flow of electrons in a circuit. Inside a copper wire, the electrons flow between and around the copper atoms, and the wire loses energy as heat when the electrons crash into the atoms. Obviously, something suppresses that process in superconductors, since the electrons flowing through them never flag. In fact, the current can flow forever as long as the superconductor remains chilled, a property first detected in mercury at −450°F in 1911. For decades, most scientists assumed that superconducting electrons simply had more space to maneuver: atoms in superconductors have much less energy to vibrate back and forth, giving electrons a wider shoulder to slip by and avoid crashes. That explanation's true as far as it goes. But really, as three scientists figured out in 1957, it's electrons themselves that metamorphose at low temperatures.

When zooming past atoms in a superconductor, electrons tug at the atoms' nuclei. The positive nuclei drift slightly toward the electrons, and this leaves a wake of higher-density positive charge. The higher-density charge attracts other electrons, which in a sense become paired with the first. It's not a strong coupling between electrons, more like the weak bond between argon and fluorine; that's why the coupling emerges only at low temperatures, when atoms aren't vibrating too much and knocking the electrons apart. At those low temperatures, you cannot think of electrons as isolated; they're stuck together and work in teams. And during their circuit, if one electron gets gummed up or knocks into an atom, its partners yank it through before it slows down. It's like that old illegal football formation where helmetless players locked arms and stormed down the field—a flying electron wedge. This microscopic state translates to superconductivity when billions of billions of pairs all do the same thing.

Incidentally, this explanation is known as the BCS theory of superconductivity, after the last names of the men who developed it: John Bardeen, Leon Cooper (the electron partners are called Cooper pairs), and Robert Schrieffer.* That's the same John Bardeen who coinvented the germanium transistor, won a Nobel Prize for it, and dropped his scrambled eggs on the floor when he heard the news. Bardeen dedicated himself to superconductivity after leaving Bell Labs for Illinois in 1951, and the BCS trio came up with the full theory six years on. It proved so good, so accurate, they shared the 1972 Nobel Prize in Physics for their work. This time, Bardeen commemorated the occasion by missing a press conference at his university because he couldn't figure out how to get his new (transistor-powered) electric garage door open. But when he visited Stockholm for the second time, he presented his two adult sons to the king of Sweden, just as he'd promised he would back in the fifties.

If elements are cooled below even superconducting temperatures, the atoms grow so loopy that they overlap and swallow each other up, a state called coherence. Coherence is crucial to understanding that impossible Einsteinian state of matter promised earlier in this chapter. Understanding coherence requires a short but thankfully element-rich detour into the nature of light and another once impossible innovation, lasers.

Few things delight the odd aesthetic sense of a physicist as much as the ambiguity, the two-in-oneness, of light. We normally think of light as waves. In fact, Einstein formulated his special theory of relativity in part by thinking about how the universe would appear to him—what space would look like, how time would (or wouldn't) pass—if he rode sidesaddle on one of those waves. (Don't ask me how he imagined this.) At the same time, Einstein proved (he's ubiquitous in this arena) that light sometimes acts like particle BBs called photons. Combining the wave and particle views (called wave-particle duality), he correctly deduced that light is not only the fastest thing in the universe, it's the fastest possible thing, at 186,000 miles per second, in a vacuum. Whether you detect light as a wave or photons depends on how you measure it, since light is neither wholly one nor the other.

Despite its austere beauty in a vacuum, light gets corrupted when it interacts with some elements. Sodium can slow light down to 38 miles per hour, almost twenty times slower than sound. Praseodymium can even catch light, hold on to it for a few seconds like a baseball, then toss it in a different direction.

Lasers manipulate light in subtler ways. Remember that electrons are like elevators: they never rise from level 1 to level 3.5 or drop from level 5 to level 1.8. Electrons jump only between whole-number levels. When excited electrons crash back down, they jettison excess energy as light, and because electron movement is so constrained, so too is the color of the light produced. It's monochromatic—at least in theory. In practice, electrons in different atoms are simultaneously dropping from level 3 to 1, or 4 to 2, or whatever—and every different drop produces a different color. Plus, different atoms emit light at different times. To our eyes, this light looks uniform, but on a photon level, it's uncoordinated and jumbled.

Lasers circumvent that timing problem by limiting what floors the elevator stops at (as do their cousins, masers, which work the same way but produce non-visible light). The most powerful, most impressive lasers today—capable of producing beams that, for a fraction of a second, produce more power than the whole United States—use crystals of yttrium spiked with neodymium. Inside the laser, a strobe light curls around the neodymium-yttrium crystal and flashes incredibly quickly at incredibly high intensities. This infusion of light excites the electrons in the neodymium and makes them jump way, way higher than normal. To keep with our elevator bit, they might rocket up to the tenth floor. Suffering vertigo, they immediately ride back down to the safety of, say, the second floor. Unlike normal crashes, though, the electrons are so disturbed that they have a breakdown and don't release their excess energy as light; they shake and release it as heat. Also, relieved at being on the safe second floor, they get off the elevator, dawdle, and don't bother hurrying down to the ground floor.

In fact, before they can hurry down, the strobe light flashes again. This sends more of the neodymium's electrons flying up to the tenth floor and crashing back down. When this happens repeatedly, the second floor gets crowded; when there are more electrons on the second floor than the first, the laser has achieved "population inversion." At this point, if any dawdling electrons do jump to the ground floor, they disturb their already skittish and crowded neighbors and knock them over the balcony, which in turn knocks others down. And notice the simple beauty of this: when the neodymium electrons drop this time, they're all dropping from two to one at the same time, so they all produce the same color of light. This coherence is the key to a laser. The rest of the laser apparatus cleans up light rays and hones the beams by bouncing them back and forth between two mirrors. But at that point, the neodymium-yttrium crystal has done its work to produce coherent, concentrated light, beams so powerful they can induce thermonuclear fusion, yet so focused they can sculpt a cornea without frying the rest of the eye.

Based on that technological description, lasers may seem more engineering challenges than scientific marvels, yet lasers—and masers, which historically came first—encountered real scientific prejudice when they were developed in the 1950s. Charles Townes remembers that even after he'd built the first working maser, senior scientists would look at him wearily and say, Sorry, Charles, that's impossible. And these weren't hacks—small-minded naysayers who lacked the imagination to see the Next Big Thing. Both John von Neumann, who helped design the basic architecture of modern computers (and modern nuclear bombs), and Niels Bohr, who did as much to explain quantum mechanics as anyone, dismissed Townes's maser to his face as simply "not possible."

Bohr and von Neumann blew it for a simple reason: they forgot about the duality of light. More specifically, the famous uncertainty principle of quantum mechanics led them astray. Because Werner Heisenberg's uncertainty principle is so easy to misunderstand—but once understood is a powerful tool for making new forms of matter—the next section will unpack this little riddle about the universe.

If nothing tickles physicists like the dual nature of light, nothing makes physicists wince like hearing someone expound on the uncertainty principle in cases where it doesn't apply. Despite what you may have heard, it has (almost*) nothing to do with observers changing things by the mere act of observing. All the principle says, in its entirety, is this:

$$\Delta x \, \Delta p \geq \frac{h}{4\pi}$$

That's it.

Now, if you translate quantum mechanics into English (always risky), the equation says that the uncertainty in something's position (Δx) times the uncertainty in its speed and direction (its momentum, Δp) always exceeds or is equal to the number "h divided by four times pi." (The h stands for Planck's constant, which is such a small number, about 100 trillion trillion times smaller than one, that the uncertainty principle applies only to tiny, tiny things such as electrons or photons.) In other words, if you know a particle's position very well, you cannot know its momentum well at all, and vice versa.

Note that these uncertainties aren't uncertainties about measuring things, as if you had a bad ruler; they're uncertainties built into nature itself. Remember how light has a reversible nature, part wave, part particle? In dismissing the laser, Bohr and von Neumann got stuck on the ways light acts like particles, or photons. To their ears, lasers sounded so precise and focused that the uncertainty in the photons' positions would be nil. That meant the uncertainty in the momentum had to be large, which meant the photons could be flying off at any energy or in any direction, which seemed to contradict the idea of a tightly focused beam.

They forgot that light behaves like waves, too, and that the rules for waves are different. For one, how can you tell where a wave is? By its nature, it spreads out—a built-in source of uncertainty. And unlike particles, waves can swallow and combine with other waves. Two rocks thrown into a pond will kick up the highest crests in the area between them, which receives energy from smaller waves on both sides.

In the laser's case, there aren't two but trillions of trillions of "rocks" (i.e., electrons) kicking up waves of light, which all mix together. The key point is that the uncertainty principle doesn't apply to sets of particles, only to individual particles.

Within a beam, a set of light particles, it's impossible to say where any one photon is located. And with such a high uncertainty about each photon's position inside the beam, you can channel its energy and direction very, very precisely and make a laser. This loophole is difficult to exploit but is enormously powerful once you've got your fingers inside it—which is why *Time* magazine honored Townes by naming him one of its "Men of the Year" (along with Pauling and Segrè) in 1960, and why Townes won a Nobel Prize in 1964 for his maser work.

In fact, scientists soon realized that much more fit inside the loophole than photons. Just as light beams have a dual particle/wave nature, the farther you burrow down and parse electrons and protons and other supposed hard particles, the fuzzier they seem. Matter, at its deepest, most enigmatic quantum level, is indeterminate and wavelike. And because, deep down, the uncertainty principle is a mathematical statement about the limits of drawing boundaries around waves, those particles fall under the aegis of uncertainty, too.

Now again, this works only on minute scales, scales where h, Planck's constant, a number 100 trillion trillion times smaller than one, isn't considered small. What embarrasses physicists is when people extrapolate up and out to human beings and claim that $\Delta x \Delta p \geq h/4\pi$ really "proves" you cannot observe something in the everyday world without changing it—or, for the heuristically daring, that objectivity itself is a scam and that scientists fool themselves into thinking they "know" anything. In truth, there's about only one case where uncertainty on a nanoscale affects anything on our macroscale: that outlandish state of matter—Bose-Einstein condensate (BEC)—promised earlier in this chapter.

This story starts in the early 1920s when Satyendra Nath Bose, a chubby, bespectacled Indian physicist, made an error while working through some quantum mechanics equations during a lecture. It was a sloppy, undergraduate boner, but it intrigued Bose. Unaware of his mistake at first, he'd worked everything out, only to find that the "wrong" answers produced by his mistake agreed very well with experiments on the properties of photons—much better than the "correct" theory.*

So as physicists have done throughout history, Bose decided to pretend that his error was the truth, admit that he didn't know why, and write a paper. His seeming mistake, plus his obscurity as an Indian, led every established scientific journal in Europe to reject it. Undaunted, Bose sent his paper directly to Albert Einstein. Einstein studied it closely and determined that Bose's answer was clever—it basically said that certain particles, like photons, could collapse on top of each other until they were indistinguishable. Einstein cleaned the paper up a little, translated it into German, and then expanded Bose's work into another, separate paper that covered not just photons but whole atoms. Using his celebrity pull, Einstein had both papers published jointly.

In them, Einstein included a few lines pointing out that if atoms got cold enough—billions of times colder than even superconductors—they would condense into a new state of matter. However, the ability to produce atoms that cold so outpaced the technology of the day that not even far-thinking Einstein could comprehend the possibility. He considered his condensate a frivolous curiosity. Amazingly, scientists got a glimpse of Bose-Einstein matter a decade later, in a type of superfluid helium where small pockets of atoms bound themselves together. The Cooper pairs of electrons in superconductors also behave like the BEC in a way. But this binding together in superfluids and superconductors was limited, and not at all like the state Einstein envisioned—his was a cold, sparse mist. Regardless, the helium and BCS people never pursued Einstein's conjecture, and nothing more happened with the BEC until 1995, when two clever scientists at the University of Colorado conjured some up with a gas of rubidium atoms.

Fittingly, one technical achievement that made real BEC possible was the laser—which was based on ideas first espoused by Bose about photons. That may seem backward, since lasers usually heat things up. But lasers can cool atoms, too, if wielded properly. On a fundamental, nanoscopic level, temperature just measures the average speed of particles. Hot molecules are furious little clashing fists, and cold molecules drag along. So the key to cooling something down is slowing its particles down. In laser cooling, scientists cross a few beams, Ghostbusters-like, and create a trap of "optical molasses." When the rubidium atoms in the gas hurtled through the molasses, the lasers pinged them with low-intensity photons. The rubidium atoms were bigger and more powerful, so this was like shooting a machine gun at a screaming asteroid. Size disparities notwithstanding, shooting an asteroid with enough bullets will eventually halt it, and that's exactly what happened to the rubidium atoms. After absorbing photons from all sides, they slowed, and slowed, and slowed some more, and their temperature dropped to just 1/10,000 of a degree above absolute zero.

Still, even that temperature is far too sweltering for the BEC (you can grasp now why Einstein was so pessimistic). So the Colorado duo, Eric Cornell and Carl Wieman, incorporated a second phase of cooling in which a magnet repeatedly sucked off the "hottest" remaining atoms in the rubidium gas. This is basically a sophisticated way of blowing on a spoonful of soup—cooling something down by pushing away warmer atoms. With the energetic atoms gone, the overall temperature kept sinking. By doing this slowly and whisking away only the few hottest atoms each time, the scientists plunged the temperature to a billionth of a degree (0.000000001) above absolute zero. At this point, finally, the sample of two thousand rubidium atoms collapsed into the Bose-Einstein condensate, the coldest, gooeyest, and most fragile mass the universe has ever known.

But to say "two thousand rubidium atoms" obscures what's so special about the BEC. There weren't two thousand rubidium atoms as much as one giant marshmallow of a rubidium atom. It was a singularity, and the explanation for why relates back to the uncertainty principle. Again, temperature just measures the average speed of atoms. If the molecules' temperature dips below a billionth of a degree, that's not much speed at all—meaning the uncertainty about that speed is absurdly low. It's basically zero. And because of the wavelike nature of atoms on that level, the uncertainty about their position must be quite large.

So large that, as the two scientists relentlessly cooled the rubidium atoms and squeezed them together, the atoms began to swell, distend, overlap, and finally disappear into each other. This left behind one large ghostly "atom" that, in theory (if it weren't so fragile), might be capacious enough to see under a microscope. That's why we can say that in this case, unlike anywhere else, the uncertainty principle has swooped upward and affected something (almost) human-sized. It took less than $100,000 worth of equipment to create this new state of matter, and the BEC held together for only ten seconds before combusting. But it held on long enough to earn Cornell and Wieman the 2001 Nobel Prize.*

As technology keeps improving, scientists have gotten better and better at inducing matter to form the BEC. It's not like anyone's taking orders yet, but scientists might soon be able to build "matter lasers" that shoot out ultra-focused beams of atoms thousands of times more powerful than light lasers, or construct "supersolid" ice cubes that can flow through each other without losing their solidity. In our sci-fi future, such things could prove every bit as amazing as light lasers and superfluids have in our own pretty remarkable age.

17

Spheres of Splendor:
The Science of Bubbles

Not every breakthrough in periodic-table science has to delve into exotic and intricate states of matter like the BEC. Everyday liquids, solids, and gases still yield secrets now and then, if fortune and the scientific muses collude in the right way. According to legend, as a matter of fact, one of the most important pieces of scientific equipment in history was invented not only *over* a glass of beer but *by* a glass of beer.

Donald Glaser—a lowly, thirsty, twenty-five-year-old junior faculty member who frequented bars near the University of Michigan—was staring one night at the bubbles streaming through his lager, and he naturally started thinking particle physics. At the time, 1952, scientists were using knowledge from the Manhattan Project and nuclear science to conjure up exotic and fragile species of particles such as kaons, muons, and pions, ghostly brothers of familiar protons, neutrons, and electrons. Particle physicists suspected, even hoped, that those particles would overthrow the periodic table as the fundamental map of matter, since they'd be able to peer even deeper into subatomic caves.

But to progress further, they needed a better way to "see" those infinitesimal particles and track how they behaved. Over his beer, Glaser—who had short, wavy hair, glasses, and a high forehead—decided bubbles were the answer. Bubbles in liquids form around imperfections or incongruities. Microscopic scratches in a champagne glass are one place they form; dissolved pockets of carbon dioxide in beer are another. As a physicist, Glaser knew that bubbles are especially prone to form as liquids heat up and approach their boiling point (think of a pan of water on the stove). In fact, if you hold a liquid just below its boiling point, it will burst into bubbles if anything agitates it.

This was a good start but still basic physics. What made Glaser stand out were the next mental steps he took. Those rare kaons, muons, and pions appear only when an atom's nucleus, its dense core, is splintered. In 1952, a device called a cloud chamber existed, in which a "gun" shot ultra-fast atomic torpedoes at cold gas atoms. Muons and kaons and so on sometimes appeared in the chamber after direct strikes, and the gas condensed into liquid drops along the particles' track. But substituting a liquid for the gas made more sense, Glaser thought. Liquids are thousands of times denser than gases, so aiming the atomic gun at, say, liquid hydrogen would cause far more collisions. Plus, if liquid hydrogen was held a shade below its boiling point, even a little kick of energy from a ghostly particle would lather up the hydrogen like Glaser's beer. Glaser also suspected he could photograph the bubble trails and then measure how different particles left different trails or spirals, depending on their size and charge.... By the time he swallowed the final bubble in his own glass, the story goes, Glaser had the whole thing worked out.

It's a story of serendipity that scientists have long wanted to believe. But sadly, like most legends, it's not entirely accurate. Glaser did invent the bubble chamber, but through careful

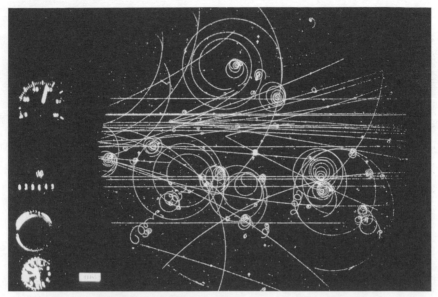

Depending on their size and charge, different subatomic particles make different swirls and spirals as they blast through a bubble chamber. The tracks are actually finely spaced bubbles in a frigid bath of liquid hydrogen. (Courtesy of CERN)

experimentation in a lab, not on a pub napkin. Happily, though, the truth is even stranger than the legend. Glaser designed his bubble chamber to work as explained above, but with one modification.

For Lord knows what reason—perhaps lingering undergraduate fascination—this young man decided beer, not hydrogen, was the best liquid to shoot the atomic gun at. He really thought that beer would lead to an epochal breakthrough in subatomic science. You can almost imagine him smuggling Budweiser into the lab at night, perhaps splitting a six-pack between science and his stomach as he filled thimble-sized beakers with America's finest, heated them almost to boiling, and bombarded them to produce the most exotic particles then known to physics.

Unfortunately for science, Glaser later said, the beer experiments flopped. Nor did lab partners appreciate the stink of vaporized ale. Undaunted, Glaser refined his experiments, and his colleague Luis Alvarez—of dinosaur-killing-asteroid fame—eventually determined the most sensible liquid to use was in fact hydrogen. Liquid hydrogen boils at –435°F, so even minute amounts of heat will make a froth. As the simplest element, hydrogen also avoided the messy complications that other elements (or beer) might cause when particles collided. Glaser's revamped "bubble chamber" provided so many insights so quickly that in 1960 he appeared among the fifteen "Men of the Year" in *Time* magazine with Linus Pauling, William Shockley, and Emilio Segrè. He also won the Nobel Prize at the disgustingly young age of thirty-three. Having moved on to Berkeley by then, he borrowed Edwin McMillan and Segrè's white vest for the ceremony.

Bubbles aren't usually counted as an essential scientific tool. Despite—or maybe because of—their ubiquity in nature and the ease of producing them, they were dismissed as toys for centuries. But when physics emerged as the dominant science in the 1900s, physicists suddenly found a lot of work for these toys in probing the most basic structures in the universe. Now that biology is ascendant, biologists use bubbles to study the development of cells, the most complex structures in the universe. Bubbles have proved to be wonderful natural laboratories for experiments in all fields, and the recent history of science can be read in parallel with the study of these "spheres of splendor."

One element that readily forms bubbles—as well as foam, a state where bubbles overlap and lose their spherical shape—is calcium. Cells are to tissues what bubbles are to foams, and the best example of a foam structure in the body (besides saliva) is spongy bone. We usually think of foams as no sturdier than shaving cream, but when certain air-infused substances dry out or cool down, they harden and stiffen, like durable versions of bath suds. NASA actually uses special foams to protect space shuttles on reentry, and calcium-enriched bones are similarly strong yet light. What's more, sculptors for millennia have carved tombstones and obelisks and false gods from pliable yet sturdy calcium rocks such as marble and limestone. These rocks form when tiny sea creatures die and their calcium-rich shells sink and pile up on the ocean floor. Like bones, shells have natural pores, but calcium's chemistry enhances their supple strength. Most natural water, such as rainwater, is slightly acidic, while calcium's minerals are slightly basic. When water leaks into calcium's pores, the two react like a mini grade-school volcano to release small amounts of carbon dioxide, which softens up the rock. On a large and geological scale, reactions between rainwater and calcium form the huge cavities we know as caves.

Beyond anatomy and art, calcium bubbles have shaped world economics and empires. The many calcium-rich coves along the southern coast of England aren't natural, but originated as limestone quarries around 55 bc, when the limestone-loving Romans arrived. Scouts sent out by Julius Caesar spotted an attractive, cream-colored limestone near modern-day Beer, England, and began chipping it out to adorn Roman facades. English limestone from Beer later was used in building Buckingham Palace, the Tower of London, and Westminster Abbey, and all that missing stone left gaping caverns in the seaside cliffs. By 1800, a few local boys who'd grown up sailing ships and playing tag in the labyrinths decided to marry their childhood pastimes by becoming smugglers, using the calcium coves to conceal the French brandy, fiddles, tobacco, and silk they ran over from Normandy in fast cutters.

The smugglers (or, as they styled themselves, free traders) thrived because of the hateful taxes the English government levied on French goods to spite Napoleon, and the scarcity of the taxed items created, inevitably, a demand bubble. Among many other things, the inability of His Majesty's expensive coast guard to crack down on smuggling convinced Parliament to liberalize trade laws in the 1840s—which brought about real free trade, and with it the economic prosperity that allowed Great Britain to expand its never-darkening empire.

Given all this history, you'd expect a long tradition of bubble science, but no. Notable minds like Benjamin Franklin (who discovered why oil calms frothy water) and Robert Boyle (who experimented on and even liked to taste the fresh, frothy urine in his chamber pot) did dabble in bubbles. And primitive physiologists sometimes did things such as bubbling gases into the blood of half-living, half-dissected dogs. But scientists mostly ignored bubbles themselves, their structure and form, and left the study of bubbles to fields that they scorned as intellectually inferior—what might be called "intuitive sciences." Intuitive sciences aren't pathological, merely fields such as horse breeding or gardening that investigate natural phenomena but that long relied more on hunches and almanacs than controlled experiments. The intuitive science that picked up bubbles research was cooking. Bakers and brewers had long used yeasts—primitive bubble-making machines—to leaven bread and carbonate beer. But eighteenth-century haute cuisine chefs in Europe learned to whip egg whites into vast, fluffy foams and began to experiment with the meringues, porous cheeses, whipped creams, and cappuccinos we love today.

Still, chefs and chemists tended to distrust one another, chemists seeing cooks as undisciplined and unscientific, cooks seeing chemists as sterile killjoys. Only around 1900 did bubble science coalesce into a respectable field, though the men responsible, Ernest Rutherford and Lord Kelvin, had only dim ideas of what their work would lead to. Rutherford, in fact, was mostly interested in plumbing what at the time were the murky depths of the periodic table.

Shortly after moving from New Zealand to Cambridge University in 1895, Rutherford devoted himself to radioactivity, the genetics or nanotechnology of the day. Natural vigorousness led Rutherford to experimental science, for he wasn't exactly a clean-fingernails guy. Having grown up hunting quail and digging potatoes on a family farm, he recalled feeling like "an ass in lion's skin" among the robed dons of Cambridge. He wore a walrus mustache, toted radioactive samples around in his pockets, and smoked foul cigars and pipes. He was given to blurting out both weird euphemisms—perhaps his devout Christian wife discouraged him from swearing—and also the bluest curses in the lab, because he couldn't help himself from damning his equipment to hell when it didn't behave. Perhaps to make up for his cursing, he also sang, loudly and quite offkey, "Onward, Christian Soldiers" as he marched around his dim lab. Despite that ogre-like description, Rutherford's outstanding scientific trait was elegance. Nobody was better, possibly in the history of science, at coaxing nature's secrets out of physical apparatus. And there's no better example than the elegance he used to solve the mystery of how one element can transform into another.

After moving from Cambridge to Montreal, Rutherford grew interested in how radioactive substances contaminate the air around them with more radioactivity. To investigate this, Rutherford built on the work of Marie Curie, but the New Zealand hick proved cagier than his more celebrated female contemporary. According to Curie (among others), radioactive elements leaked a sort of gas of "pure radioactivity" that charged the air, just as lightbulbs flood the air with light. Rutherford suspected that "pure radioactivity" was actually an unknown gaseous element with its own radioactive properties. As a result, whereas Curie spent months boiling down thousands of pounds of black, bubbling pitchblende to get microscopic samples of radium and polonium, Rutherford sensed a shortcut and let nature work for him. He simply let an active sample decay in a closed container, then drew bubbles of the gas off into an inverted flask, which gave him all the radioactive material he needed. Rutherford and his collaborator, Frederick Soddy, quickly proved the radioactive bubbles were in fact a new element, radon. And because the sample beneath the beaker shrank in proportion as the radon sample grew in volume, they realized that one element actually mutated into another.

Not only did Rutherford and Soddy find a new element, they discovered novel rules for jumping around on the periodic table. Elements could suddenly move laterally as they decayed and skip across spaces. This was thrilling but blasphemous. Science had finally discredited and excommunicated the chemical magicians who'd claimed to turn lead into gold, and here Rutherford and Soddy were opening the gate back up. When Soddy finally let himself believe what was happening and burst out, "Rutherford, this is transmutation!" Rutherford had a fit.

"For Mike's sake, Soddy," he boomed. "Don't call it transmutation. They'll have our heads off as alchemists!"

The radon sample soon midwifed even more startling science. Rutherford had arbitrarily named the little bits that flew off radioactive atoms alpha particles. (He also discovered beta particles.) Based on the weight differences between generations of decaying elements, Rutherford suspected that alphas were actually helium atoms breaking off and escaping like bubbles through a boiling liquid. If this was true, elements could do more than hop two spaces on the periodic table like pieces on a typical board game; if uranium emitted helium, elements were jumping from one side of the table to the other like a lucky (or disastrous) move in Snakes & Ladders.

To test this idea, Rutherford had his physics department's glassblowers blow two bulbs. One was soap-bubble thin, and he pumped radon into it. The other was thicker and wider, and it surrounded the first. The alpha particles had enough energy to tunnel through the first glass shell but not the second, so they became trapped in the vacuum cavity between them. After a few days, this wasn't much of an experiment, since the trapped alpha particles were colorless and didn't really do anything. But then Rutherford ran a battery current through the cavity. If you've ever traveled to Tokyo or New York, you know what happened. Like all noble gases, helium glows when excited by electricity, and Rutherford's mystery particles began glowing helium's characteristic green and yellow. Rutherford basically proved that alpha particles were escaped helium atoms with an early "neon" light. It was a perfect example of his elegance, and also his belief in dramatic science.

With typical flair, Rutherford announced the alpha-helium connection during his acceptance speech for the 1908 Nobel Prize. (In addition to winning the prize himself, Rutherford mentored and hand-trained eleven future prizewinners, the last in 1978, more than four decades after Rutherford died. It was perhaps the most impressive feat of progeny since Genghis Khan fathered hundreds of children seven centuries earlier.) His findings intoxicated the Nobel audience. Nevertheless, the most immediate and practical application of Rutherford's helium work probably escaped many in Stockholm. As a consummate experimentalist, however, Rutherford knew that truly great research didn't just support or disprove a given theory, but fathered more experiments. In particular, the alpha-helium experiment allowed him to pick the scab off the old theological-scientific debate about the true age of the earth.

The first semi-defensible guess for that age came in 1650, when Irish archbishop James Ussher worked backward from "data" such as the begats list in the Bible ("...and Serug lived thirty years, and begat Nahor...and Nahor lived nine and twenty years, and begat Terah," etc.) and calculated that God had finally gotten around to creating the earth on October 23, 4004 BC. Ussher did the best he could with the available evidence, but within decades that date was proved laughably late by most every scientific field. Physicists could even pin precise numbers on their guesses by using the equations of thermodynamics. Just as hot coffee cools down in a freezer, physicists knew that the earth constantly loses heat to space, which is cold. By measuring the rate of lost heat and extrapolating backward to when every rock on earth was molten, they could estimate the earth's date of origin. The premier scientist of the nineteenth century, William Thomson, known as Lord Kelvin, spent decades on this problem and in the late 1800s announced that the earth had been born twenty million years before.

It was a triumph of human reasoning—and about as dead wrong as Ussher's guess. By 1900, Rutherford among others recognized that however far physics had outpaced other sciences in prestige and glamour (Rutherford himself was fond of saying, "In science, there is *only* physics; all the rest is stamp collecting"—words he later had to eat when he won a Nobel Prize in Chemistry), in this case the physics didn't feel right. Charles Darwin argued persuasively that humans could not have evolved from dumb bacteria in just twenty million years, and followers of Scottish geologist James Hutton argued that no mountains or canyons could have formed in so short a span. But no one could unravel Lord Kelvin's formidable calculations until Rutherford started poking around in uranium rocks for bubbles of helium.

Inside certain rocks, uranium atoms spit out alpha particles (which have two protons) and transmutate into element ninety, thorium. Thorium then begets radium by spitting out another alpha particle. Radium begets radon with yet another, and radon begets polonium, and polonium begets stable lead. This was a well-known deterioration. But in a stroke of genius akin to Glaser's, Rutherford realized that those alpha particles, after being ejected, form small bubbles of helium inside rocks. The key insight was that helium never reacts with or is attracted to other elements. So unlike carbon dioxide in limestone, helium shouldn't normally be inside rocks. Any helium that *is* inside rocks was therefore fathered by radioactive decay. Lots of helium inside a rock means that it's old, while scant traces indicate it's a youngster.

Rutherford had thought about this process for a few years by 1904, when he was thirty-three and Kelvin was eighty. By that age, despite all that Kelvin had contributed to science, his mind had fogged. Gone were the days when he could put forward exciting new theories, like the one that all the elements on the periodic table were, at their deepest levels, twisted "knots of ether" of different shapes. Most detrimentally to his science, Kelvin never could incorporate the unsettling, even frightening science of radioactivity into his worldview. (That's why Marie Curie once pulled him, too, into a closet to look at her glow-in-the-dark element—to instruct him.) In contrast, Rutherford realized that radioactivity in the earth's crust would generate extra heat, which would bollix the old man's theories about a simple heat loss into space.

Excited to present his ideas, Rutherford arranged a lecture in Cambridge. But however dotty Kelvin got, he was still a force in scientific politics, and demolishing the old man's proudest calculation could in turn jeopardize Rutherford's career. Rutherford began the speech warily, but luckily, just after he started, Kelvin nodded off in the front row. Rutherford raced to get to his conclusions, but just as he began knocking the knees out from under Kelvin's work, the old man sat up, refreshed and bright.

Trapped onstage, Rutherford suddenly remembered a throwaway line he'd read in Kelvin's work. It said, in typically couched scientific language, that Kelvin's calculations about the earth's age were correct *unless someone discovered extra sources of heat* inside the earth. Rutherford mentioned that qualification, pointed out that radioactivity might be that latent source, and with masterly spin ad-libbed that Kelvin had therefore predicted the discovery of radioactivity dozens of years earlier. What genius! The old man glanced around the audience, radiant. He thought that Rutherford was full of crap, but he wasn't about to disregard the compliment.

Rutherford laid low until Kelvin died, in 1907, then he soon proved the helium-uranium connection. And with no politics stopping him now—in fact, he became an eminent peer himself (and later ended up as scientific royalty, too, with a box on the periodic table, element 104, rutherfordium)—the eventual Lord Rutherford got some primordial uranium rock, eluted the helium from microscopic bubbles inside, and determined that the earth was at least 500 million years old—twenty-five times greater than Kelvin's guess and the first calculation correct to within a factor of ten. Within years, geologists with more experience finessing rocks took over for Rutherford and determined that the helium pockets proved the earth to be at least two billion years old. This number was still 50 percent too low, but thanks to the tiny, inert bubbles inside radioactive rocks, human beings at last began to face the astounding age of the cosmos.

After Rutherford, digging for small bubbles of elements inside rocks became standard work in geology. One especially fruitful approach uses zircon, a mineral that contains zirconium, the pawnshop heartbreaker and knockoff jewelry substitute.

For chemical reasons, zircons are hardy—zirconium sits below titanium on the periodic table and makes convincing fake diamonds for a reason. Unlike soft rocks such as limestone, many zircons have survived since the early years of the earth, often as hard, poppy-seed grains inside larger rocks. Due to their unique chemistry, when zircon crystals formed way back when, they vacuumed up stray uranium and packed it into atomic bubbles inside themselves. At the same time, zircons had a distaste for lead and squeezed that element out (the opposite of what meteors do). Of course, that didn't last long, since uranium decays into lead, but the zircons had trouble working the lead slivers out again. As a result, any lead inside lead-phobic zircons nowadays has to be a daughter product of uranium. The story should be familiar by now: after measuring the ratio of lead to uranium in zircons, it's just a matter of graphing backward to year zero. Anytime you hear scientists announcing a record for the "world's oldest rock"—probably in Australia or Greenland, where zircons have survived the longest—rest assured they used zircon-uranium bubbles to date it.

Other fields adopted bubbles as a paradigm, too. Glaser began experimenting with his bubble chamber in the 1950s, and around that same time, theoretical physicists such as John Archibald Wheeler began speaking of the universe as foam on its fundamental level. On that scale, billions of trillions of times smaller than atoms, Wheeler dreamed that "the glassy smooth spacetime of the atomic and particle worlds gives way.... There would literally be no left and right, no before and after. Ordinary ideas of length would disappear. Ordinary ideas of time would evaporate. I can think of no better name than quantum foam for this state of affairs." Some cosmologists today calculate that our entire universe burst into existence when a single submicronanobubble slipped free from that foam and began expanding at an exponential rate. It's a handsome theory, actually, and explains a lot—except, unfortunately, why this might have happened.

Ironically, Wheeler's quantum foam traces its intellectual lineage to the ultimate physicist of the classical, everyday world, Lord Kelvin. Kelvin didn't invent froth science—that was a blind Belgian with the fitting name (considering how little influence his work had) of Joseph Plateau. But Kelvin did popularize the science by saying things like he could spend a lifetime scrutinizing a single soap bubble. That was actually disingenuous, since according to his lab notebooks, Kelvin formulated the outline of his bubble work one lazy morning in bed, and he produced just one short paper on it. Still, there are wonderful stories of this white-bearded Victorian splashing around in basins of water and glycerin, with what looked like a miniature box spring on a ladle, to make colonies of interlocking bubbles. And *squarish* bubbles at that, reminiscent of the *Peanuts* character Rerun, since the box spring's coils were shaped into rectangular prisms.

Plus, Kelvin's work gathered momentum and inspired real science in future generations. Biologist D'Arcy Wentworth Thompson applied Kelvin's theorems on bubble formation to cell development in his seminal 1917 book *On Growth and*

Form, a book once called "the finest work of literature in all the annals of science that have been recorded in the English tongue." The modern field of cell biology began at this point. What's more, recent biochemical research hints that bubbles were the efficient cause of life itself. The first complex organic molecules may have formed not in the turbulent ocean, as is commonly thought, but in water bubbles trapped in Arctic-like sheets of ice. Water is quite heavy, and when water freezes, it crushes together dissolved "impurities," such as organic molecules, inside bubbles. The concentration and compression in those bubbles might have been high enough to fuse those molecules into self-replicating systems. Furthermore, recognizing a good trick, nature has plagiarized the bubble blueprint ever since. Regardless of where the first organic molecules formed, in ice or ocean, the first crude cells were certainly bubble-like structures that surrounded proteins or RNA or DNA and protected them from being washed away or eroded. Even today, four billion years later, cells still have a basic bubble design.

Kelvin's work also inspired military science. During World War I, another lord, Lord Rayleigh, took on the urgent wartime problem of why submarine propellers were so prone to disintegrate and decay, even when the rest of the hull remained intact. It turned out that bubbles produced by the churning propellers turned around and attacked the metal blades like sugar attacks teeth, and with similarly corrosive results. Submarine science led to another breakthrough in bubble research as well—though at the time this finding seemed unpromising, even dodgy. Thanks to the memory of German U-boats, studying sonar—sound waves moving in water—was as trendy in the 1930s as radioactivity had been before. At least two research teams discovered that if they rocked a tank with jet engine–level noise, the bubbles that appeared would sometimes collapse and wink at them with a flash of blue or green light. (Think of bit-ing wintergreen Life Savers in a dark closet.) More interested in blowing up submarines, scientists didn't pursue so-called sonoluminescence, but for fifty years it hung on as a scientific parlor trick, passed down from generation to generation.

It might have remained just that if not for a colleague taunting Seth Putterman one day in the mid-1980s. Putter-man worked at the University of California at Los Angeles in fluid dynamics, a fiendishly tricky field. In some sense, sci-entists know more about distant galaxies than about turbu-lent water gushing through sewer pipes. The colleague was teasing Putterman about this ignorance, when he mentioned that Putterman's ilk couldn't even explain how sound waves can transmutate bubbles into light. Putterman thought that sounded like an urban legend. But after looking up the scant research that existed on sonoluminescence, he chucked his pre-vious work to study blinking bubbles full-time.*

For Putterman's first, delightfully low-tech experiments, he set a beaker of water between two stereo speakers, which were cranked to dog-whistle frequencies. A heated toaster wire in the beaker kicked up bubbles, and sound waves trapped and levitated them in the water. Then came the fun part. Sound waves vary between barren, low-intensity troughs and crests of high intensity. The tiny, trapped bubbles responded to low pressure by swelling a thousand times, like a balloon filling a room. After the sound wave bottomed out, the high-pressure front tore in and crushed the bubble's volume by half a million times, at forces 100 billion times greater than gravity. Not sur-prisingly, it's that supernova crush that produces the eerie light. Most amazingly, despite being squished into a "singularity," a term rarely used outside the study of black holes, the bubble stays intact. After the pressure lifts, the bubble billows out again, unpopped, as if nothing had happened. It's then squished again and blinks again, with the process repeating thousands of times every second.

Putterman soon bought more sophisticated equipment than his original garage-band setup, and upon doing so, he had a run-in with the periodic table. To help determine what exactly caused the bubbles to sparkle, he began trying different gases. He found that although bubbles of plain air produced nice crackles of blue and green, pure nitrogen or oxygen, which together make up 99 percent of air, wouldn't luminesce, no matter what volume or shrillness he cranked the sound to. Perturbed, Putterman began pumping trace gases from air back into the bubbles until he found the elemental flint—argon.

That was odd, since argon is an inert gas. What's more, the only other gases Putterman (and a growing cadre of bubble scientists) could get to work were argon's heavier chemical cousins, krypton and especially xenon. In fact, when rocked with sonar, xenon and krypton flared up even brighter than argon, producing "stars in a jar" that sizzled at 35,000°F inside water— far hotter than the surface of the sun. Again, this was baffling. Xenon and krypton are often used in industry to smother fires or runaway reactions, and there was no reason to think those dull, inert gases could produce such intense bubbles.

Unless, that is, their inertness is a covert asset. Oxygen, carbon dioxide, and other atmospheric gases inside bubbles can use the incoming sonar energy to divide or react with one another. From the point of view of sonoluminescence, that's energy squandered. Some scientists, though, think that inert gases under high pressure cannot help but soak up sonar energy. And with no way to dissipate the energy, bubbles of xenon or krypton collapse and have no choice but to propagate and concentrate energy in the bubbles' cores. If that's the case, then the noble gases' nonreactivity is the key to sonoluminescence. Whatever the reason, the link to sonoluminescence will rewrite what it means to be an inert gas.

Unfortunately, tempted by harnessing that high energy, some scientists (including Putterman) have linked this fragile bubble science with desktop fusion, a cousin of that all-time favorite pathological science. (Because of the temperatures involved, it's not cold fusion.) There has long been a vague, free-association link between bubbles and fusion, partly because Boris Deryagin, an influential Soviet scientist who studied the stability of foams, believed strongly in cold fusion. (Once, in an inconceivable experiment, the antithesis of one of Rutherford's, Deryagin supposedly tried to induce cold fusion in water by firing a Kalashnikov rifle into it.)

The dubious link between sonoluminescence and fusion (sonofusion) was made explicit in 2002 when the journal *Science* ran a radioactively controversial paper on sonoluminescence-driven nuclear power. Unusually, *Science* also ran an editorial admitting that many senior scientists thought the paper flawed if not fraudulent; even Putterman recommended that the journal reject this one. *Science* printed it anyway (perhaps so everyone would have to buy a copy to see what all the fuss was about). The paper's lead author was later hauled before the U.S. House of Representatives for faking data.

Thankfully, bubble science had a strong enough foundation* to survive that disgrace. Physicists interested in alternative energy now model superconductors with bubbles. Pathologists describe AIDS as a "foamy" virus, for the way infected cells swell before exploding. Entomologists know of insects that use bubbles like submersibles to breathe underwater, and ornithologists know that the metallic sheen of peacocks' plumage comes from light tickling bubbles in the feathers. Most important, in 2008, in food science, students at Appalachian State University finally determined what makes Diet Coke explode when you drop Mentos into it. Bubbles. The grainy surface of Mentos candy acts as a net to snag small dissolved bubbles, which are stitched into large ones. Eventually, a few gigantic bubbles break off, rocket upward, and whoosh through the nozzle, spurting up to twenty magnificent feet. This discovery was undoubtedly the greatest moment in bubble science since Donald Glaser eyed his lager more than fifty years before and dreamed of subverting the periodic table.

18

Tools of Ridiculous Precision

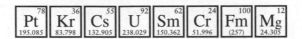

Think of the most fussy science teacher you ever had. The one who docked your grade if the sixth decimal place in your answer was rounded incorrectly; who tucked in his periodic table T-shirt, corrected every student who said "weight" when he or she meant "mass," and made everyone, including himself, wear goggles even while mixing sugar water. Now try to imagine someone whom your teacher would hate for being anal-retentive. *That* is the kind of person who works for a bureau of standards and measurement.

Most countries have a standards bureau, whose job it is to measure *everything*—from how long a second really is to how much mercury you can safely consume in bovine livers (very little, according to the U.S. National Institute of Standards and Technology, or NIST). To scientists who work at standards bureaus, measurement isn't just a practice that makes science possible; it's a science in itself. Progress in any number of fields, from post-Einsteinian cosmology to the astrobiological hunt for life on other planets, depends on our ability to make ever finer measurements based on ever smaller scraps of information.

For historical reasons (the French Enlightenment folk were fanatic measurers), the Bureau International des Poids

*The two-inch-wide International Prototype Kilogram (center), made of plati-
num and iridium, spends all day every day beneath three nested bell jars inside a
humidity- and temperature-controlled vault in Paris. Surrounding the Kilogram
are six official copies, each under two bell jars. (Reproduced with permission of
BIPM, which retains full international protected copyright)*

et Mesures (BIPM) just outside Paris acts as the standards
bureau's standards bureau, making sure all the "franchises" stay
in line. One of the more peculiar jobs of the BIPM is coddling
the International Prototype Kilogram—the world's official
kilogram. It's a two-inch-wide, 90 percent platinum cylinder
that, by definition, has a mass of exactly 1.000000...kilogram

(to as many decimal places as you like). I'd say that's about two pounds, but I'd feel guilty about being inexact.

Because the Kilogram is a physical object and therefore damageable, and because the definition of a kilogram ought to stay constant, the BIPM must make sure it never gets scratched, never attracts a speck of dust, never loses (the bureau hopes!) a single atom. For if any of that happened, its mass could spike to 1.000000...1 kilograms or plummet to 0.999999...9 kilograms, and the mere possibility induces ulcers in a national bureau of standards type. So, like phobic mothers, they constantly monitor the Kilogram's temperature and the pressure around it to prevent microscopic bloating and contracting, stress that could slough off atoms. It's also swaddled within three successively smaller bell jars to prevent humidity from condensing on the surface and leaving a nanoscale film. And the Kilogram is made from dense platinum (and iridium) to minimize the surface area exposed to unacceptably dirty air, the kind we breathe. Platinum also conducts electricity well, which cuts down on the buildup of "parasitic" static electricity (the BIPM's word) that might zap stray atoms.

Finally, platinum's toughness mitigates against the chance of a disastrous fingernail nick on the rare occasions when people actually lay a hand on the Kilogram. Other countries need their own official 1.000000...cylinder to avoid having to fly to Paris every time they want to measure something precisely, and since the Kilogram is *the* standard, each country's knockoff has to be compared against it. The United States has had its official kilogram, called K20 (i.e., the twentieth official copy), which resides in a government building in exurban Maryland, calibrated just once since 2000, and it's due for another calibration, says Zeina Jabbour, group leader for the NIST mass and force team. But calibration is a multimonth process, and security regulations since 2001 have made flying K20 to Paris an absolute hassle. "We have to hand-carry the kilograms through the flight," says Jabbour, "and it's hard to get through security and customs with a slug of metal, and tell people they cannot touch it." Even opening K20's customized suitcase in a "dusty airport" could compromise it, she says, "and if somebody insists on touching it, that's the end of the calibration."

Usually, the BIPM uses one of six official copies of the Kilogram (each kept under two bell jars) to calibrate the knockoffs. But the official copies have to be measured against their own standard, so every few years scientists remove the Kilogram from its vault (using tongs and wearing latex gloves, of course, so as not to leave fingerprints—but not the powdery kind of gloves, because that would leave a residue—oh, and not holding it for too long, because the person's body temperature could heat it up and ruin everything) and calibrate the calibrators.* Alarmingly, scientists noticed during calibrations in the 1990s that, even accounting for atoms that rub off when people touch it, in the past few decades the Kilogram had lost an additional mass equal to that of a fingerprint(!), half a microgram per year. No one knows why.

The failure—and it is that—to keep the Kilogram perfectly constant has renewed discussions about the ultimate dream of every scientist who obsesses over the cylinder: to make it obsolete. Science owes much of its progress since about 1600 to adopting, whenever possible, an objective, non-human-centered point of view about the universe. (This is called the Copernican principle, or less flatteringly the mediocrity principle.) The kilogram is one of seven "base units" of measurement that permeate all branches of science, and it's no longer acceptable for any of those units to be based on a human artifact, especially if it's mysteriously shrinking.

The goal with every unit, as England's bureau of national standards cheekily puts it, is for one scientist to e-mail its definition to a colleague on another continent and for the colleague to be able to reproduce something with exactly those dimensions, based only on the description in the e-mail. You can't e-mail the Kilogram, and no one has ever come up with a definition more reliable than that squat, shiny, pampered cylinder in Paris. (Scientists are drawing closer, but so far the best ideas are either too impossibly involved—such as counting trillions of trillions of atoms—or requires measurements too precise for even the best instruments today.) The inability to solve the kilogram conundrum, to either stop it from shrinking or superannuate it, has become an increasing source of international worry and embarrassment (at least for us anal types).

The pain is all the more acute because the kilogram is the last base unit bound to human strictures. A platinum rod in Paris defined 1.000000...meter through much of the twentieth century, until scientists redefined it with a krypton atom in 1960, fixing it at 1,650,763.73 wavelengths of red-orange light from a krypton-86 atom. This distance is virtually identical to the length of the old rod, but it made the rod obsolete, since that many wavelengths of krypton light would stretch the same distance in any vacuum anywhere. (*That's* an e-mailable definition.) Since then, measurement scientists (metrologists) have re-redefined a meter (about three feet) as the distance any light travels in a vacuum in 1/299,792,458 of a second.

Similarly, the official definition of one second used to be about 1/86,400 of one spin on Earth's axis (i.e., the number of seconds in one day). But a few pesky facts made that an inconvenient standard. More important, the length of a day is slowly increasing because of the sloshing of ocean tides, which drag and slow earth's rotation. To correct for this, metrologists slip in a "leap second" about every third year, usually when no one's paying attention, at midnight on December 31. But leap seconds are an ugly, ad hoc solution. And rather than tie a supposedly universal unit of time to the transit of an unremarkable rock around a forgettable star, the U.S. standards bureau has developed cesium-based atomic clocks.

Atomic clocks run on the same leaping and crashing of excited electrons we've discussed before. But atomic clocks also exploit a subtler movement, the electrons' "fine structure." If the normal jump of an electron resembles a singer jumping an octave from G to G, fine structure resembles a jump from G to G-flat or G-sharp. Fine structure effects are most noticeable in magnetic fields, and they're caused by things you can safely ignore unless you find yourself in a dense, high-level physics course—such as the magnetic interactions between electrons and protons or corrections due to Einstein's relativity. The upshot is that after those fine adjustments,* each electron jumps either slightly lower (G-flat) or slightly higher (G-sharp) than expected.

The electron "decides" which jump to make based on its intrinsic spin, so one electron never hits the sharp and the flat on successive leaps. It hits one or the other every time. Inside atomic clocks, which look like tall, skinny pneumatic tubes, a magnet purges all the cesium atoms whose outer electrons jump to one level, call it G-flat. That leaves only atoms with G-sharp electrons, which are gathered into a chamber and excited by an intense microwave. This causes cesium electrons to pop (i.e., jump and crash) and emit photons of light. Each cycle of jumping up and down is elastic and always takes the same (extremely short) amount of time, so the atomic clock can measure time simply by counting photons. Really, whether you purge the G-flat or G-sharp doesn't matter, but you have to purge one of them because jumping to either level takes a different amount of time, and at the scales metrologists work with, such imprecision is unacceptable.

Cesium proved convenient as the mainspring for atomic clocks because it has one electron exposed in its outermost shell, with no nearby electrons to muffle it. Cesium's heavy, lumbering atoms are fat targets for the maser that strums them as well. Still, even in plodding cesium, the outer electron is a quick bugger. Instead of a few dozen or few thousand times per second, it performs 9,192,631,770 back-and-forths every one-Mississippi. Scientists picked that ungainly number instead of cutting themselves off at 9,192,631,769 or letting things drag on until 9,192,631,771 because it matched their best guess for a second back in 1955, when they built the first cesium clock. Regardless, 9,192,631,770 is now fixed. It became the first base-unit definition to achieve universal e-mailability, and it even helped liberate the meter from its platinum rod after 1960.

Scientists adopted the cesium standard as the world's official measurement of time in the 1960s, replacing the astronomical second, and while the cesium standard has profited science by ensuring precision and accuracy worldwide, humanity has undeniably lost something. Since before even the ancient Egyptians and Babylonians, human beings used the stars and seasons to track time and record their most important moments. Cesium severed that link with the heavens, effaced it just as surely as urban streetlamps blot out constellations. However fine an element, cesium lacks the mythic feeling of the moon or sun. Besides, even the argument for switching to cesium—its universality, since cesium electrons should vibrate at the same frequency in every pocket of the universe—may no longer be a safe bet.

*   *   *

If anything runs deeper than a mathematician's love of variables, it's a scientist's love of constants. The charge of the electron, the strength of gravity, the speed of light—no matter the experiment, no matter the circumstances, those parameters never vary. If they did, scientists would have to chuck the precision that separates "hard" sciences from social sciences like economics, where whims and sheer human idiocy make universal laws impossible.

Even more seductive to scientists, because more abstract and universal, are fundamental constants. Obviously, the numerical value of a particle's size or speed would change if we arbitrarily decided that meters should be longer or if the kilogram suddenly shrank (ahem). Fundamental constants, however, don't depend on measurement. Like π, they're pure, fixed numbers, and also like π, they pop up in all sorts of contexts that seem tantalizingly explainable but that have so far resisted all explanation.

The best-known dimensionless constant is the fine structure constant, which is related to the fine splitting of electrons. In short, it controls how tightly negative electrons are bound to the positive nucleus. It also determines the strength of some nuclear processes. In fact, if the fine structure constant— which I'll refer to as alpha, because that's what scientists call it—if alpha had been slightly smaller right after the big bang, nuclear fusion in stars would never have gotten hot enough to fuse carbon. Conversely, if alpha had grown slightly larger, carbon atoms would all have disintegrated aeons ago, long before finding their way into us. That alpha avoided this atomic Scylla and Charybdis makes scientists thankful, naturally, but also very antsy, because they cannot explain how it succeeded. Even a good, inveterate atheist like physicist Richard Feynman once said of the fine structure constant, "All good theoretical physicists put this number up on their wall and worry about it....It's one of the greatest damn mysteries of physics: a magic number that comes to us with no understanding by man. You might say the 'hand of God' wrote that number, and we don't know how He pushed His pencil."

Historically, that didn't stop people from trying to decipher this scientific *mene, mene, tekel, upharsin*. English astronomer Arthur Eddington, who during a solar eclipse in 1919 provided the first experimental proof of Einstein's relativity, grew fascinated with alpha. Eddington had a penchant, and it must be said a talent, for numerology,* and in the early 1900s, after alpha was measured to be around 1/136, Eddington began concocting "proofs" that alpha equaled exactly 1/136, partly because he found a mathematical link between 136 and 666. (One colleague derisively suggested rewriting the book of Revelation to take this "finding" into account.) Later measurements showed that alpha was closer to 1/137, but Eddington just tossed a 1 into his formula somewhere and continued on as if his sand castle hadn't crumbled (earning him the immortal nickname Sir Arthur Adding-One). A friend who later ran across Eddington in a cloakroom in Stockholm was chagrined to see that he insisted on hanging his hat on peg 137.

Today alpha equals 1/137.0359 or so. Regardless, its value makes the periodic table possible. It allows atoms to exist and also allows them to react with sufficient vigor to form compounds, since electrons neither roam too freely from their nuclei nor cling too closely. This just-right balance has led many scientists to conclude that the universe couldn't have hit upon its fine structure constant by accident. Theologians, being more explicit, say alpha proves that a creator has "programmed" the universe to produce both molecules and, possibly, life. That's why it was such a big deal in 1976 when a Soviet (now American) scientist named Alexander Shlyakhter scrutinized a bizarre site in Africa called Oklo and declared that alpha, a fundamental and invariant constant of the universe, was getting bigger.

Oklo is a galactic marvel: the only *natural* nuclear fission reactor known to exist. It stirred to life some 1.7 billion years ago, and when French miners unearthed the dormant site in 1972, it caused a scientific roar. Some scientists argued that Oklo couldn't have happened, while some fringe groups pounced on Oklo as "evidence" for outlandish pet theories such as long-lost African civilizations and crash landings by nuclear-powered alien star cruisers. Actually, as nuclear scientists determined, Oklo was powered by nothing but uranium, water, and blue-green algae (i.e., pond scum). Really. Algae in a river near Oklo produced excess oxygen after undergoing photosynthesis. The oxygen made the water so acidic that as it trickled underground through loose soil, it dissolved the uranium from the bedrock. All uranium back then had a higher concentration of the bomb-ready uranium-235 isotope—about 3 percent, compared to 0.7 percent today. So the water was volatile already, and when underground algae filtered the water, the uranium was concentrated in one spot, achieving a critical mass.

Though necessary, a critical mass wasn't sufficient. In general, for a chain reaction to occur, uranium nuclei must not only be struck by neutrons, they must absorb them. When pure uranium fissions, its atoms shoot out "fast" neutrons that bounce off neighbors like stones skipped across water. Those are basically duds, wasted neutrons. Oklo uranium went nuclear only because the river water slowed the neutrons down enough for neighboring nuclei to snag them. Without the water, the reaction never would have begun.

But there's more. Fission also produces heat, obviously. And the reason there's not a big crater in Africa today is that when the uranium got hot, it boiled the water away. With no water, the neutrons became too fast to absorb, and the process ground to a halt. Only when the uranium cooled down did water trickle back in—which slowed the neutrons and restarted the reactor. It was a nuclear Old Faithful, self-regulating, and it consumed 13,000 pounds of uranium over 150,000 years at sixteen sites around Oklo, in on/off cycles of 150 minutes.

How did scientists piece that tale together 1.7 billion years later? With elements. Elements are mixed thoroughly in the earth's crust, so the ratios of different isotopes should be the same everywhere. At Oklo, the uranium-235 concentration was 0.003 to 0.3 percent less than normal—a huge difference. But what determined that Oklo was a natural nuke and not the remnants of a smuggling operation for rogue terrorists was the overabundance of useless elements such as neodymium. Neodymium mostly comes in three even-numbered flavors, 142, 144, and 146. Uranium fission reactors produce odd-numbered neodymium at higher rates than normal. In fact, when scientists analyzed the neodymium concentrations at Oklo and subtracted out the natural neodymium, they found that Oklo's nuclear "signature" matched that of a modern, man-made fission reactor. Amazing.

Still, if neodymium matched, other elements didn't. When Shlyakhter compared the Oklo nuclear waste to modern waste in 1976, he found that too little of some types of samarium had formed. By itself, that's not so thrilling. But again, nuclear processes are reproducible to a stunning degree; elements such as samarium don't just fail to form. So samarium's digression hinted to Shlyakhter that something had been off back then. Taking a hell of a leap, he calculated that if only the fine structure constant had been just a fraction smaller when Oklo went nuclear, the discrepancies were easy to explain. In this, he resembled the Indian physicist Bose, who didn't claim to know why his "wrong" equations about photons explained so much;

he only knew they did. The problem was, alpha is a fundamental constant. It *can't* vary, not according to physics. Worse for some, if alpha varied, probably no one (or, rather, no One) had "tuned" alpha to produce life after all.

With so much at stake, many scientists since 1976 have reinterpreted and challenged the alpha-Oklo link. The changes they're measuring are so small and the geological record so piecemeal after 1.7 billion years, it seems unlikely anyone will ever prove anything definitive about alpha from Oklo data. But again, never underestimate the value of throwing an idea out there. Shlyakhter's samarium work whetted the appetite of dozens of ambitious physicists who wanted to knock off old theories, and the study of changing constants is now an active field. One boost to these scientists was the realization that even if alpha has changed very little since "only" 1.7 billion years ago, it might have shifted rapidly during the first billion years of the universe, a time of primordial chaos. As a matter of fact, after investigating star systems called quasars and interstellar dust clouds, some Australian astronomers* claim they've detected the first real evidence of inconstants.

Quasars are black holes that tear apart and cannibalize other stars, violence that releases gobs and gobs of light energy. Of course, when astronomers collect that light, they're not looking at events in real time, but events that took place long, long ago, since light takes time to cross the universe. What the Australians did was examine how huge storms of interstellar space dust affected the passage of ancient quasar light. When light passes through a dust cloud, vaporized elements in the cloud absorb it. But unlike something opaque, which absorbs all light, the elements in the cloud absorb light at specific frequencies. Moreover, similar to atomic clocks, elements absorb light not of one narrow color but of two very finely split colors.

The Australians had little luck with some elements in the dust clouds; it turns out those elements would hardly notice if alpha vacillated every day. So they expanded their search to elements such as chromium, which proved highly sensitive to alpha: the smaller alpha was in the past, the redder the light that chromium absorbed and the narrower the spaces between its G-flat and G-sharp levels. By analyzing the gap that chromium and other elements produced billions of years ago near the quasar and comparing it with atoms in the lab today, scientists can judge whether alpha has changed in the meantime. And though, like all scientists—especially ones proposing something controversial—the Australians hedge and couch their findings in scientific language about such and such only "being consistent with the hypothesis" of this and that, they do think their ultrafine measurements indicate that alpha changed by up to 0.001 percent over ten billion years.

Now, honestly, that may seem a ridiculous amount to squabble over, like Bill Gates fighting for pennies on the sidewalk. But the magnitude is less important than the *possibility* of a fundamental constant changing.* Many scientists dispute the results from Australia, but if those results hold up—or if any of the other scientists working on variable constants find proof positive—scientists would have to rethink the big bang, because the only laws of the universe they know would not quite have held from the beginning.* A variable alpha would overthrow Einsteinian physics in the same way Einstein deposed Newton and Newton deposed medieval Scholastic physics. And as the next section shows, a drifting alpha might also revolutionize how scientists explore the cosmos for signs of life.

We've already met Enrico Fermi in rather poor circumstances— he died of beryllium poisoning after some brash experiments and won a Nobel Prize for discovering transuranic elements he didn't discover. But it isn't right just to leave you with a negative impression of this dynamo. Scientists loved Fermi universally and without reserve. He's the namesake of element one hundred, fermium, and he's regarded as the last great dual-purpose, theoretical-cum-experimental scientist, someone equally likely to have grease from laboratory machines on his hands as chalk from the blackboard. He had a devilishly quick mind as well. During scientific meetings with colleagues, they sometimes needed to run to their offices to look up arcane equations to resolve some point; often as not when they returned, Fermi, unable to wait, had derived the entire equation from scratch and had the answer they needed. Once, he asked junior colleagues to figure out how many millimeters thick the dust could get on the famously dirty windows in his lab before the dust avalanched under its own weight and sloughed onto the floor. History doesn't record the answer, only the impish* question.

Not even Fermi, however, could wrap his head around one hauntingly simple question. As noted earlier, many philosophers marvel that the universe seems fine-tuned to produce life because certain fundamental constants have a "perfect" value. Moreover, scientists have long believed—in the same spirit they believe a second shouldn't be based on our planet's orbit—that earth is not cosmically special. Given that ordinariness, as well as the immense numbers of stars and planets, and the aeons that have passed since the big bang (and leaving aside any sticky religious issues), the universe should rightfully be swarming with life. Yet not only have we never met alien creatures, we've never even gotten a hello. As Fermi brooded on those contradictory facts over lunch one day, he cried out to his colleagues, as if he expected an answer, "Then where is everybody?"

His colleagues burst out laughing at what's now known as

"Fermi's paradox." But other scientists took Fermi seriously, and they really believed they could get at an answer. The best-known attempt came in 1961, when astrophysicist Frank Drake laid out what's now known as the Drake Equation. Like the uncertainty principle, the Drake Equation has had a layer of interpretation laid over it that obscures what it really says. In short, it's a series of guesses: about how many stars exist in the galaxy, what fraction of those have earthlike planets, what fraction of those planets have intelligent life, what fraction of those life forms would want to make contact, and so on. Drake originally calculated* that ten sociable civilizations existed in our galaxy. But again, that was just an informed guess, which led many scientists to renounce it as flatulent philosophizing. How on earth, for instance, can we psychoanalyze aliens and figure out what percent want to chat?

Nonetheless, the Drake Equation is important: it outlines what data astronomers need to collect, and it put astrobiology on a scientific foundation. Perhaps someday we'll look back on it as we do early attempts to organize a periodic table. And with vast improvements recently in telescopes and other heavenly measuring devices, astrobiologists have tools to provide more than guesses. In fact, the Hubble Space Telescope and others have teased out so much information from so little data that astrobiologists can now go one better than Drake. They don't have to wait for intelligent alien life to seek us out or even scour deep space for an alien Great Wall of China. They might be able to measure direct evidence of life—even mute life such as exotic plants or festering microbes—by searching for elements such as magnesium.

Obviously, magnesium is less important than oxygen or carbon, but element twelve could be a huge help for primitive creatures, allowing them to transition from organic molecules to real life. Almost all life forms use metallic elements in trace amounts to create, store, or shuttle energetic molecules around inside them. Animals primarily use the iron in hemoglobin, but the earliest and most successful forms of life, especially blue-green algae, used magnesium. Specifically, chlorophyll (probably the most important organic chemical on earth—it drives photosynthesis by converting stellar energy into sugars, the basis of the food chain) is crowned with magnesium ions at its center. Magnesium in animals helps DNA function properly.

Magnesium deposits on planets also imply the presence of liquid H_2O, the most probable medium for life to arise in. Magnesium compounds sponge up water, so even on bare, rocky planets like Mars, there's hope of finding bacteria (or bacterial fossils) among those kinds of deposits. On watery planets (like a great candidate for extraterrestrial life in our solar system, Jupiter's moon Europa), magnesium helps keep oceans fluid. Europa has an icy outer crust, but huge liquid oceans thrive beneath it, and satellite evidence indicates that those oceans are full of magnesium salts. Like any dissolved substances, magnesium salts depress the freezing point of water so that it stays liquid at lower temperatures. Magnesium salts also stir "brine volcanism" on the rocky floors beneath the oceans. Those salts swell the volume of water they're dissolved in, and the extra pressure from the extra volume powers volcanoes that spew brackish water and churn the ocean depths. (The pressure also cracks surface ice caps, spilling rich ice into the ocean water—which is good, in case intra-ice bubbles are important in creating life.) Moreover, magnesium compounds (among others) can provide the raw materials to build life by eroding carbon-rich chemicals from the ocean floor. Short of landing a probe or seeing alien vegetation, detecting magnesium salts on a bare, airless planet is a good sign something might be happening there bio-wise.

But let's say Europa is barren. Even though the hunt for

far-flung alien life has grown more technologically sophisticated, it still rests on one huge assumption: that the same science that controls us locally holds true in other galaxies and held true at other times. But if alpha changed over time, the consequences for potential alien life could be enormous. Historically, life perhaps couldn't exist until alpha "relaxed" enough to allow stable carbon atoms to form—and perhaps then life arose effortlessly, without any need to appeal to a creator. And because Einstein determined that space and time are intertwined, some physicists believe that alpha variations in time could imply alpha variations across space. According to this theory, just as life arose on earth and not the moon because earth has water and an atmosphere, perhaps life arose here, on a seemingly random planet in a seemingly unremarkable pocket of space, because only here do the proper cosmological conditions exist for sturdy atoms and full molecules. This would resolve Fermi's paradox in a cinch: nobody has come calling because nobody's there.

At this moment, the evidence leans toward the ordinariness of earth. And based on the gravitational perturbations of far-off stars, astronomers now know of thousands of planets, which makes the odds of finding life somewhere quite good. Still, the great debate of astrobiology will be deciding whether earth, and by extension human beings, have a privileged place in the universe. Hunting for alien life will take every bit of measuring genius we have, possibly with some overlooked box on the periodic table. All we know for sure is that if some astronomer turned a telescope to a far-off star cluster tonight and found incontrovertible evidence of life, even microbial scavengers, it would be the most important discovery ever—proof that human beings are not so special after all. Except that we exist, too, and can understand and make such discoveries.

19

Above (and Beyond)
the Periodic Table

There's a conundrum near the fringes of the periodic table. Highly radioactive elements are always scarce, so you'd think, intuitively, the element that falls apart the most easily would also be the most scarce. And the element that is effaced most quickly and thoroughly whenever it appears in the earth's crust, ultra-fragile francium, is indeed rare. Francium winks out of existence on a timescale quicker than any other natural atom—yet one element is even rarer than francium. It's a paradox, and resolving the paradox actually requires leaving behind the comfortable confines of the periodic table. It requires setting out for what nuclear physicists consider their New World, their America to conquer—the "island of stability"—which is their best and perhaps only hope for extending the table beyond its current limitations.

As we know, 90 percent of particles in the universe are hydrogen, and the other 10 percent are helium. Everything else, including six million billion billion kilos of earth, is a cosmic rounding error. And in that six million billion billion kilos, the total amount of astatine, the scarcest natural element, is one stupid ounce. To put that into some sort of (barely) comprehensible scope, imagine that you lost your Buick Astatine in an immense parking garage and you have zero idea where it is. Imagine the tedium of walking down every row on every level past every space, looking for your vehicle. To mimic hunting for astatine atoms inside the earth, that parking garage would have to be about 100 million spaces wide, have 100 million rows, and be 100 million stories high. And there would have to be 160 identical garages just as big—and in all those buildings, there'd be just one Astatine. You'd be better off walking home.

If astatine is so rare, it's natural to ask how scientists ever took a census of it. The answer is, they cheated a little. Any astatine present in the early earth has long since disintegrated radioactively, but other radioactive elements sometimes decay into astatine after they spit out alpha or beta particles. By knowing the total amount of the parent elements (usually elements near uranium) and calculating the odds that each of those will decay into astatine, scientists can ring up some plausible numbers for how many astatine atoms exist. This works for other elements, too. For instance, at least twenty to thirty ounces of astatine's near neighbor on the periodic table, francium, exist at any moment.

Funnily enough, astatine is at the same time far more robust than francium. If you had a million atoms of the longest-lived type of astatine, half of them would disintegrate in four hundred minutes. A similar sample of francium would hang on for just twenty minutes. Francium is so fragile it's basically useless, and even though there's (barely) enough of it in the earth for chemists to detect it directly, no one will ever herd enough atoms of it together to make a visible sample. If they did, it would be so intensely radioactive it would murder them immediately. (The current flash-mob record for francium is ten thousand atoms.)

No one will likely ever produce a visible sample of astatine either, but at least it's good for something—as a quick-acting radioisotope in medicine. In fact, after scientists—led by our old friend Emilio Segrè—identified astatine in 1939, they injected a sample into a guinea pig to study it. Because astatine sits below iodine on the periodic table, it acts like iodine in the body and so was selectively filtered and concentrated by the rodent's thyroid gland. Astatine remains the only element whose discovery was confirmed by a nonprimate.

The odd reciprocity between astatine and francium begins in their nuclei. There, as in all atoms, two forces struggle for dominance: the strong nuclear force (which is always attractive) and the electrostatic force (which can repel particles). Though the most powerful of nature's four fundamental forces, the strong nuclear force has ridiculously short arms. Think *Tyrannosaurus rex*. If particles stray more than a few trillionths of an inch apart, the strong force is impotent. For that reason, it rarely comes into play outside nuclei and black holes. Yet within its range, it's a hundred times more muscular than the electrostatic force. That's good, because it keeps protons and neutrons bound together instead of letting the electrostatic force wrench nuclei apart.

When you get to nuclei the size of astatine and francium, the limited reach really catches up with the strong force, and it has trouble binding all the protons and neutrons together. Francium has eighty-seven protons, none of which want to touch. Its 130-odd neutrons buffer the positive charges well but also add so much bulk that the strong force cannot reach all the way across a nucleus to quell civil strife. This makes francium (and astatine, for similar reasons) highly unstable. And it stands to reason that adding more protons would increase electric repulsion, making atoms heavier than francium even weaker.

That's only sort of correct, though. Remember that Maria Goeppert-Mayer ("S.D. Mother Wins Nobel Prize") developed a theory about long-lived "magic" elements—atoms with two, eight, twenty, twenty-eight, etc., protons or neutrons that were extra-stable. Other numbers of protons or neutrons, such as ninety-two, also form compact and fairly stable nuclei, where the short-leashed strong force can tighten its grip on protons. That's why uranium is more stable than either astatine or francium, despite being heavier. As you move down the periodic table element by element, then, the struggle between the strong nuclear and electrostatic forces resembles a plummeting stock market ticker, with an overall downward trend in stability, but with many wiggles and fluctuations as one force gains the upper hand, then the other.*

Based on this prevailing pattern, scientists assumed that the elements beyond uranium would asymptotically approach a life span of 0.0. But as they groped forward with the ultra-heavy elements in the 1950s and 1960s, something unexpected happened. In theory, magic numbers extend until infinity, and it turned out that there was a quasi-stable nucleus after uranium, at element 114. And instead of it being fractionally more stable, scientists at (where else?) the University of California at Berkeley calculated that 114 might survive orders of magnitude longer than the ten or so heavy elements preceding it. Given the dismally short life span of heavy elements (microseconds at best), this was a wild, counterintuitive idea. Packing neutrons and protons onto most man-made elements is like packing on explosives, since you're putting more stress on the nucleus. Yet with element 114, packing on more TNT seemed to *steady* the bomb. Just as strangely, elements such as 112 and 116 seemed (on paper at least) to get horseshoes-and-kisses benefits from having close to 114 protons. Even being around that quasi-

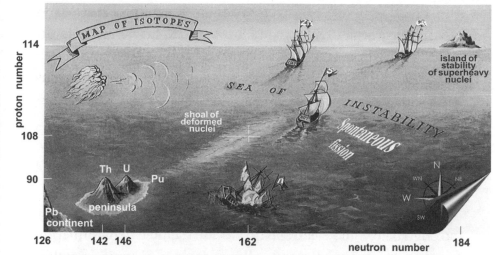

A whimsical map of the fabled "island of stability," a clump of superheavy elements that scientists hope will allow them to extend the periodic table far past its present bounds. Notice the stable lead (Pb) continent of the main-body periodic table, the watery trench of unstable elements, and the small, semi-stable peaks at thorium and uranium before the sea opens up. (Yuri Oganessian, Joint Institute for Nuclear Research, Dubna, Russia)

magic number calmed them. Scientists began calling this cluster of elements the island of stability.

Charmed by their own metaphor, and flattering themselves as brave explorers, scientists began preparing to conquer the island. They spoke of finding an elemental "Atlantis," and some, like old-time sailors, even produced sepia "charts" of unknown nucleic seas. (You'd half expect to see krakens drawn in the waters.) And for decades now, attempts to reach that oasis of superheavy elements have made up one of the most exciting fields of physics. Scientists haven't reached land yet (to get truly stable, doubly magic elements, they need to figure out ways to add more neutrons to their targets), but they're in the island's shallows, paddling around for a harbor.

Of course, an island of stability implies a stretch of

submerged stability—a stretch centered on francium. Element eighty-seven is stranded between a magic nucleus at eighty-two and a quasi-stable nucleus at ninety-two, and it's all too tempting for its neutrons and protons to abandon ship and swim. In fact, because of the poor structural foundation of its nucleus, francium is not only the least stable natural element, it's less stable than every synthetic element up to 104, the ungainly rutherfordium. If there's a "trench of instability," francium is gargling bubbles at the bottom of the Mariana.

Still, it's more abundant than astatine. Why? Because many radioactive elements around uranium happen to decay into francium as they disintegrate. But francium, instead of doing the normal alpha decay and thereby converting itself (through the loss of two protons) into astatine, decides more than 99.9 percent of the time to relieve the pressure in its nucleus by undergoing beta decay and becoming radium. Radium then undergoes a cascade of alpha decays that leap over astatine. In other words, the path of many decaying atoms leads to a short layover on francium—hence the twenty to thirty ounces of it. At the same time, francium shuttles atoms away from astatine, causing astatine to remain rare. Conundrum solved.

Now that we've plumbed the trenches, what about that island of stability? It's doubtful that chemists will ever synthesize all the elements up to very high magic numbers. But perhaps they can synthesize a stable element 114, then 126, then go from there. Some scientists believe, too, that adding electrons to extra-heavy atoms can stabilize their nuclei—the electrons might act as springs and shocks to absorb the energy that atoms normally dedicate to tearing themselves apart. If that's so, maybe elements in the 140s, 160s, and 180s are possible. The island of stability would become a chain of islands. These stable islands would get farther apart, but perhaps, like

Polynesian canoers, scientists can cross some wild distances on the new periodic archipelago.

The thrilling part is that those new elements, instead of being just heavier versions of what we already know, could have novel properties (remember how lead emerges from a lineage of carbon and silicon). According to some calculations, if electrons can tame superheavy nuclei and make them more stable, those nuclei can manipulate electrons, too—in which case, electrons might fill the atoms' shells and orbitals in a different order. Elements whose address on the table should make them normal heavy metals might fill in their octets early and act like metallic noble gases instead.

Not to tempt the gods of hubris, but scientists already have names for those hypothetical elements. You may have noticed that the extra-heavy elements along the bottom of the table get three letters instead of two and that all of them start with *u*. Once again, it's the lingering influence of Latin and Greek. As yet undiscovered element 119, Uue, is un·un·ennium; element 122, Ubb, is un·bi·bium;* and so on. Those elements will receive "real" names if they're ever made, but for now scientists can jot them down—and mark off other elements of interest, such as magic number 184, un·oct·quadium—with Latin substitutes. (And thank goodness for them. With the impending death of the binomial species system in biology—the system that gave us *Felis catus* for the house cat is gradually being replaced with chromosomal DNA "bar codes," so good-bye *Homo sapiens*, the knowing ape, hello TCATCGGTCATTGG...—the *u* elements remain about the only holdouts of once-dominant Latin in science.*)

So how far can this island-hopping extend? Can we watch little volcanoes rise beneath the periodic table forever, watch it expand and stretch down to the fittingly wide Eee, enn·enn·ennium, element 999, or even beyond? *Sigh*, no. Even if scientists figure out how to glue extra-heavy elements together, and even if they land smack on the farther-off islands of stability, they'll almost certainly skid right off into the messy seas.

The reason traces back to Albert Einstein and the biggest failure of his career. Despite the earnest belief of most of his fans, Einstein did not win his Nobel Prize for the theory of relativity, special or general. He won for explaining a strange effect in quantum mechanics, the photoelectric effect. His solution provided the first real evidence that quantum mechanics wasn't a crude stopgap for justifying anomalous experiments, but actually corresponds to reality. And the fact that Einstein came up with it is ironic for two reasons. One, as he got older and crustier, Einstein came to distrust quantum mechanics. Its statistical and deeply probabilistic nature sounded too much like gambling to him, and it prompted him to object that "God does not play dice with the universe." He was wrong, and it's too bad that most people have never heard the rejoinder by Niels Bohr: "Einstein! Stop telling God what to do."

Second, although Einstein spent his career trying to unify quantum mechanics and relativity into a coherent and svelte "theory of everything," he failed. Not completely, however. Sometimes when the two theories touch, they complement each other brilliantly: relativistic corrections of the speed of electrons help explain why mercury (the element I'm always looking out for) is a liquid and not the expected solid at room temperature. And no one could have created his namesake element, number ninety-nine, einsteinium, without knowledge of both theories. But overall, Einstein's ideas on gravity, the speed of light, and relativity don't quite fit with quantum mechanics. In some cases where the two theories come into contact, such as inside black holes, all the fancy equations break down.

That breakdown could set limits on the periodic table. To

return to the electron-planet analogy, just as Mercury zips around the sun every three months while Neptune drags on for 165 years, inner electrons orbit much more quickly around a nucleus than electrons in outer shells. The exact speed depends on the ratio between the number of protons present and alpha, the fine structure constant discussed last chapter. As that ratio gets closer and closer to one, electrons fly closer and closer to the speed of light. But remember that alpha is (we think) fixed at 1/137 or so. Beyond 137 protons, the inner electrons would seem to be going faster than the speed of light—which, according to Einstein's relativity theory, can never happen.

This hypothetically last element, 137, is often called "feynmanium," after Richard Feynman, the physicist who first noticed this pickle. He's also the one who called alpha "one of the great damn mysteries of the universe," and now you can see why. As the irresistible force of quantum mechanics meets the immovable object of relativity just past feynmanium, something has to give. No one knows what.

Some physicists, the kind of people who think seriously about time travel, think that relativity may have a loophole that allows special (and, conveniently, unobservable) particles called tachyons to go faster than light's 186,000 miles per second. The catch with tachyons is that they may move backward in time. So if super-chemists someday create feynmanium-plus-one, un·tri·octium, would its inner electrons become time travelers while the rest of the atom sits pat? Probably not. Probably the speed of light simply puts a hard cap on the size of atoms, which would obliterate those fanciful islands of stability as thoroughly as A-bomb tests did coral atolls in the 1950s.

So does that mean the periodic table will be kaput soon? Fixed and frozen, a fossil?

No, no, and no again.

*   *   *

If aliens ever land and park here, there's no guarantee we'll be able to communicate with them, even going beyond the obvious fact they won't speak "Earth." They might use pheromones or pulses of light instead of sounds; they might also be, especially on the off off chance they're not made of carbon, poisonous to be around. Even if we do break into their minds, our primary concerns—love, gods, respect, family, money, peace—may not register with them. About the only things we can drop in front of them and be sure they'll grasp are numbers like pi and the periodic table.

Of course, that should be the *properties* of the periodic table, since the standard castles-with-turrets look of our table, though chiseled into the back of every extant chemistry book, is just one possible arrangement of elements. Many of our grandfathers grew up with quite a different table, one just eight columns wide all the way down. It looked more like a calendar, with all the rows of the transition metals triangled off into half boxes, like those unfortunate 30s and 31s in awkwardly arranged months. Even more dubiously, a few people shoved the lanthanides into the main body of the table, creating a crowded mess.

No one thought to give the transition metals a little more space until Glenn Seaborg and his colleagues at (wait for it) the University of California at Berkeley made over the entire periodic table between the late 1930s and early 1960s. It wasn't just that they added elements. They also realized that elements like actinium didn't fit into the scheme they'd grown up with. Again, it sounds odd to say, but chemists before this didn't take periodicity seriously enough. They thought the lanthanides and their annoying chemistry were exceptions to the normal periodic table rules—that no elements below the lanthanides would ever bury electrons and deviate from transition-metal chemistry in the same way. But the lanthanide chemistry does repeat. It has to: that's the categorical imperative of chemistry, *the* property of elements the aliens would recognize. And they'd recognize as surely as Seaborg did that the elements diverge into something new and strange right after actinium, element eighty-nine.

Actinium was the key element in giving the modern periodic table its shape, since Seaborg and his colleagues decided to cleave all the heavy elements known at the time—now called the actinides, after their first brother—and cordon them off at the bottom of the table. As long as they were moving those elements, they decided to give the transition metals more elbow room, too, and instead of cramming them into triangles, they added ten columns to the table. This blueprint made so much sense that many people copied Seaborg. It took a while for the hard-liners who preferred the old table to die off, but in the 1970s the periodic calendar finally shifted to become the periodic castle, the bulwark of modern chemistry.

But who says that's the ideal shape? The columnar form has dominated since Mendeleev's day, but Mendeleev himself designed thirty different periodic tables, and by the 1970s scientists had designed more than seven hundred variations. Some chemists like to snap off the turret on one side and attach it to the other, so the periodic table looks like an awkward staircase. Others fuss with hydrogen and helium, dropping them into different columns to emphasize that those two non-octet elements get themselves into strange situations chemically.

Really, though, once you start playing around with the periodic table's form, there's no reason to limit yourself to rectilinear shapes.* One clever modern periodic table looks like a honeycomb, with each hexagonal box spiraling outward in wider and wider arms from the hydrogen core. Astronomers and astrophysicists might like the version where a hydrogen "sun" sits at the center of the table, and all the other elements orbit it like planets with moons. Biologists have mapped the periodic table onto helixes, like our DNA, and geeks have sketched out periodic tables where rows and columns double back on themselves and wrap around the paper like the board game Parcheesi. Someone even holds a U.S. patent (#6361324) for a pyramidal Rubik's Cube toy whose twistable faces contain elements.

Musically inclined people have graphed elements onto musical staffs, and our old friend William Crookes, the spiritualist seeker, designed two fittingly fanciful periodic tables, one that looked like a lute and another like a pretzel. My own favorite tables are a pyramid-shaped one—which very sensibly gets wider row by row and demonstrates graphically where new orbitals arise and how many more elements fit themselves into the overall system—and a cutout one with twists in the middle, which I can't quite figure out but enjoy because it looks like a Möbius strip.

We don't even have to limit periodic tables to two dimensions anymore. The negatively charged antiprotons that Segrè discovered in 1955 pair very nicely with antielectrons (i.e., positrons) to form anti-hydrogen atoms. In theory, every other anti-element on the anti–periodic table might exist, too. And beyond just that looking-glass version of the regular periodic table, chemists are exploring new forms of matter that could multiply the number of known "elements" into the hundreds if not thousands.

First are superatoms. These clusters—between eight and one hundred atoms of one element—have the eerie ability to mimic single atoms of different elements. For instance, thirteen aluminium atoms grouped together in the right way do a killer bromine: the two entities are indistinguishable in chemical reactions. This happens despite the cluster being thirteen times larger than a single bromine atom and despite aluminium being nothing like the lacrimatory poison-gas staple. Other combinations of aluminium can mimic noble gases, semiconductors, bone materials like calcium, or elements from pretty much any other region of the periodic table.

The clusters work like this. The atoms arrange themselves into a three-dimensional polyhedron, and each atom in it mimics a proton or neutron in a collective nucleus. The caveat is that electrons can flow around inside this soft nucleic blob, and the atoms share the electrons collectively. Scientists wryly call this state of matter "jellium." Depending on the shape of the polyhedron and the number of corners and edges, the jellium will have more or fewer electrons to farm out and react with other atoms. If it has seven, it acts like bromine or a halogen. If four, it acts like silicon or a semiconductor. Sodium atoms can also become jellium and mimic other elements. And there's no reason to think that still other elements cannot imitate other elements, or even all the elements imitate all the other elements—an utterly Borgesian mess. These discoveries are forcing scientists to construct parallel periodic tables to classify all the new species, tables that, like transparencies in an anatomy textbook, must be layered on top of the periodic skeleton.

Weird as jellium is, the clusters at least resemble normal atoms. Not so with the second way of adding depth to the periodic table. A quantum dot is a sort of holographic, virtual atom that nonetheless obeys the rules of quantum mechanics. Different elements can make quantum dots, but one of the best is indium. It's a silvery metal, a relative of aluminium, and lives just on the borderland between metals and semiconductors.

Scientists start construction of a quantum dot by building a tiny Devils Tower, barely visible to the eye. Like geologic strata, this tower consists of layers—from the bottom up, there's a semiconductor, a thin layer of an insulator (a ceramic), indium, a thicker layer of a ceramic, and a cap of metal on top. A positive charge is applied to the metal cap, which attracts electrons. They race upward until they reach the insulator, which they cannot flow through. However, if the insulator is thin enough, an electron—which at its fundamental level is just a wave—can pull some voodoo quantum mechanical stuff and "tunnel" through to the indium.

At this point, scientists snap off the voltage, trapping the orphan electron. Indium happens to be good at letting electrons flow around between atoms, but not so good that an electron disappears inside the layer. The electron sort of hovers instead, mobile but discrete, and if the indium layer is thin enough and narrow enough, the thousand or so indium atoms band together and act like one collective atom, all of them sharing the trapped electron. It's a superorganism. Put two or more electrons in the quantum dot, and they'll take on opposite spins inside the indium and separate in oversized orbitals and shells. It's hard to overstate how weird this is, like getting the giant atoms of the Bose-Einstein condensate but without all the fuss of cooling things down to billionths of a degree above absolute zero. And it isn't an idle exercise: the dots show enormous potential for next-generation "quantum computers," because scientists can control, and therefore perform calculations with, individual electrons, a much faster and cleaner procedure than channeling billions of electrons through semiconductors in Jack Kilby's fifty-year-old integrated circuits.

Nor will the periodic table be the same after quantum dots. Because the dots, also called pancake atoms, are so flat, the electron shells are different than usual. In fact, so far the pan-cake periodic table looks quite different than the periodic table we're used to. It's narrower, for one thing, since the octet rule doesn't hold. Electrons fill up shells more quickly, and nonreactive noble gases are separated by fewer elements. That doesn't stop other, more reactive quantum dots from sharing electrons and bonding with other nearby quantum dots to form...well, who knows what the hell they are. Unlike with superatoms, there aren't any real-world elements that form tidy analogues to quantum-dot "elements."

In the end, though, there's little doubt that Seaborg's table of rows and turrets, with the lanthanides and actinides like moats along the bottom, will dominate chemistry classes for generations to come. It's a good combination of easy to make and easy to learn. But it's a shame more textbook publishers don't balance Seaborg's table, which appears inside the front cover of every chemistry book, with a few of the more suggestive periodic table arrangements inside the back cover: 3D shapes that pop and buckle on the page and that bend far-distant elements near each other, sparking some link in the imagination when you finally see them side by side. I wish very much that I could donate $1,000 to some nonprofit group to support tinkering with wild new periodic tables based on whatever organizing principles people can imagine. The current periodic table has served us well so far, but reenvisioning and recreating it is important for humans (some of us, at least). Moreover, if aliens ever do descend, I want them to be impressed with our ingenuity. And maybe, just maybe, for them to see some shape they recognize among our collection.

Then again, maybe our good old boxy array of rows and turrets, and its marvelous, clean simplicity, will grab them.

And maybe, despite all their alternative arrangements of elements, and despite all they know about superatoms and quantum dots, they'll see something new in this table. Maybe as we explain how to read the table on all its different levels, they'll whistle (or whatever) in real admiration—staggered at all we human beings have managed to pack into our periodic table of the elements.

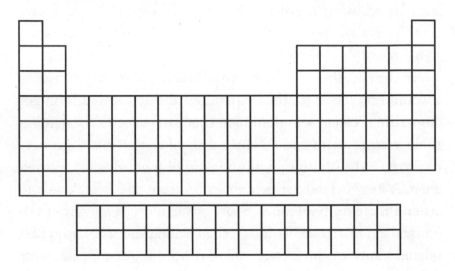

ACKNOWLEDGMENTS
AND THANKS

I would first like to thank my dear ones. My parents, who got me writing, and never asked too often what exactly I was going to do with myself once I'd started. My lovely Paula, who held my hand. My siblings, Ben and Becca, who taught me mischief. All my other friends and family from South Dakota and around the country, who supported me and got me out of the house. And finally my various teachers and professors, who first related many of the stories here, without realizing they were doing something so valuable.

I would furthermore like to thank my agent, Rick Broadhead, who believed that this project was a swell idea and that I was the one to write it. I owe a lot as well to my editor at Little, Brown, John Parsley, who saw what this book could be and helped shape it. Also invaluable were others at and around Little, Brown, including Cara Eisenpress, Sarah Murphy, Peggy Freudenthal, Barbara Jatkola, and many unnamed others who helped design and improve this book.

I offer thanks, too, to the many, many people who contributed to individual chapters and passages, either by fleshing out stories, helping me hunt down information, or offering their time to explain something to me. These include Stefan Fajans;

Theodore Gray of www.periodictable.com; Barbara Stewart at Alcoa; Jim Marshall of the University of North Texas; Eric Scerri of the University of California at Los Angeles; Chris Reed at the University of California, Riverside; Nadia Izakson; the communications team at Chemical Abstracts Service; and the staff and science reference librarians at the Library of Congress. If I've left anyone off this list, my apologies. I remain thankful, if embarrassed.

Finally, I owe a special debt of gratitude to Dmitri Mendeleev, Julius Lother Meyer, John Newlands, Alexandre-Emile Béguyer de Chancourtois, William Odling, Gustavus Hinrichs, and the other scientists who developed the periodic table—as well as thousands of other scientists who contributed to these fascinating stories about the elements.

Notes and Errata

Introduction

p. 7, "literature, poison forensics, and psychology": Another topic I learned about via mercury was meteorology. The final peal of the death knell of alchemy sounded on the day after Christmas in 1759, when two Russian scientists, trying to see how cold they could get a mixture of snow and acid, accidentally froze the quicksilver in their thermometer. This was the first recorded case of solid Hg, and with that evidence, the alchemists' immortal fluid was banished to the realm of normal matter.

Lately mercury has been politicized as well, as activists in the United States have campaigned vigorously against the (totally unfounded) dangers of mercury in vaccines.

1. Geography Is Destiny

p. 16, "anything but a pure element": Two scientists observed the first evidence for helium (an unknown spectral line, in the yellow range) during an eclipse in 1868—hence the element's name, from *helios*, Greek for "sun." The element was not isolated on earth until 1895, through the careful isolation of helium from rocks. (For more on this, see chapter 17.) For eight years, helium was thought to exist on earth in minute quantities only, until miners found a huge underground cache in Kansas in 1903. They had tried to light the gas shooting out of a vent in the ground on fire, but it wouldn't catch.

p. 18, "only the electrons matter": To reiterate the point about atoms being mostly empty space, Allan Blackman, a chemist at the University of Otago in New Zealand, wrote in the January 28, 2008, *Otago Daily Times:* "Consider the most dense known element, iridium; a sample of this the size of a tennis ball would weigh just over 3 kilograms [6.6 pounds]....Let's assume that we could somehow pack the iridium nuclei together as tight as we possibly could, thereby eliminating most of that empty space....A tennis ball–sized sample of this compacted material would now weigh an astonishing seven trillion tonnes [7.7 trillion U.S. tons]."

As a footnote to this footnote, no one really knows whether iridium is the densest element. Its density is so close to osmium's that scientists cannot distinguish between them, and in the past few decades they've traded places as king of the mountain. Osmium is on top at the moment.

p. 19, "every quibbling error": For more detailed portraits of Lewis and Nernst (and many other characters, such as Linus Pauling and Fritz Haber), I highly recommend *Cathedrals of Science: The Personalities and Rivalries That Made Modern Chemistry* by Patrick Coffey. It's a personality-driven account of the most important era in modern chemistry, between about 1890 and 1930.

p. 21, "most colorful history on the periodic table": Other facts about antimony:

1. Much of our knowledge of alchemy and antimony comes from a 1604 book, *The Triumphal Chariot of Antimony*, written by Johann Thölde. To give his book a publicity boost, Thölde claimed he'd merely translated it from a 1450 text written by a monk, Basilius Valentinus. Fearing persecution for his beliefs, Valentinus had supposedly hidden the text in a pillar in his monastery. It remained hidden until a "miraculous thunderbolt" split the pillar in Thölde's time and allowed him to discover the manuscript.

2. Although many did call antimony a hermaphrodite, others insisted it was the essence of femininity—so much so that a version of the alchemical symbol for antimony, ♀, became associated with the general symbol for "female."

3. In the 1930s in China, a poor province made do with what it had and decided to make money from antimony, about the only local resource. But antimony is soft, easily rubbed away, and slightly toxic, all of which makes for poor coins, and the government soon withdrew them. Though worth just fractions of a cent then, these coins fetch thousands of dollars from collectors today.

2. Near Twins and Black Sheep

p. 32, "really wrote Shakespeare's plays": A simpler but less colorful definition of *honorificabilitudinitatibus* is "with honorableness." The Bacon anagram for the word is "Hi ludi, F. Baconis nati, tuiti orbi," which translates to "These plays, born of F[rancis] Bacon, are preserved for the world."

p. 33, "anaconda runs 1,185 letters": There's some confusion over the longest word to appear in *Chemical Abstracts*. Many people list the tobacco mosaic virus protein, $C_{785}H_{1220}N_{212}O_{248}S_2$, but a substantial number instead list "tryptophan synthetase α protein," a relative of the chemical that people (wrongly) suppose makes them sleepy when they eat turkey (that's an urban legend). The tryptophan protein, $C_{1289}H_{2051}N_{343}O_{375}S_8$, runs 1,913 letters, over 60 percent longer than the mosaic virus protein, and numerous sources—some editions of *Guinness World Records*, the Urban Dictionary (www.urbandictionary.com), *Mrs. Byrne's Dictionary of Unusual, Obscure, and Preposterous Words*—all list tryptophan as the champ. But after spending hours in the dimly lit stacks of the Library of Congress, I never located the tryptophan molecule in *Chemical Abstracts*. It just doesn't seem to have appeared in its full, spelled-out form. To be doubly sure, I hunted down the academic paper that announced the decoding of the tryptophan protein (which was separate from the *Chemical Abstracts* listing), and there the authors chose to abbreviate the amino acid sequence. So its full name has never appeared in print as far as I can tell, which probably explains why *Guinness* later rescinded the listing for it as the longest word.

I did manage to track down listings for the mosaic virus, which is spelled out twice—first on page 967F of a brownish volume called *Chemical Abstracts Formula Index, Jan.–June 1964*, then on page 6717F of *Chemical Abstracts 7th Coll. Formulas, $C_{23}H_{32}$–Z, 56–65, 1962–1966*. Both books are compendiums that collect data for all the scholarly chemistry papers published between the dates on their covers. That means, contra other references to the world's longest word (especially on the Web), the mosaic virus listing appeared only when those tomes came out in 1964 and 1966 and not in 1972.

There's more: the tryptophan paper came out in 1964, and there are other molecules listed in that 1962–1966 *Chemical Abstracts* compendium with more Cs, Hs, Ns, Os, and Ss than the tobacco mosaic virus. So why aren't they spelled out? Because those papers appeared after 1965, the year Chemical Abstracts Service, the company in Ohio that collects all this data, overhauled its system for naming new compounds and began discouraging excessively eye-glazing names. But so why did they bother spelling out the tobacco mosaic virus protein in a 1966 compendium? It could have been chopped down but was grandfathered in. And to throw in one more twist, the original 1964 tobacco mosaic virus paper was in German. But *Chemical Abstracts* is an English-language document, in the fine reference-work tradition of Samuel Johnson and the *OED*, and it printed the name not to show off but to propagate knowledge, so it sure counts.

Whew.

By the way, I owe Eric Shively, Crystal Poole Bradley, and especially Jim Corning at Chemical Abstracts Service a lot for helping me figure all this out. They didn't have to field my confused questions ("Hi. I'm trying to find the longest word in English, and I'm not sure what it is…"), but they did.

Incidentally, on top of being the first virus discovered, the tobacco mosaic virus was the first to have its shape and structure analyzed in a rigorous way. Some of the best work in this area was done by Rosalind Franklin, the crystallography expert who generously but naively shared her data with Watson and Crick (see chapter 8). Oh, and the "α" in "tryptophan synthetase α protein" traces back to Linus Pauling's work on how proteins know how to fold into the proper shape (see chapter 8 again).

p. 36, "mercifully known as titin": A few very patient souls have posted the entire amino acid sequence of titin online. Here are the stats: It occupies forty-seven single-spaced pages of a Microsoft Word document in Times New Roman 12-point font. It contains over 34,000 amino acids, and there are 43,781 occurrences of *l*; 30,710 of *y*; 27,120 of *yl*; and just 9,229 of *e*.

p. 40, "almost a proof in itself": From a PBS *Frontline* piece called "Breast Implants on Trial": "The silicon content of living organisms decreases as the complexity of the organism rises. The ratio of silicon to carbon is 250:1 in the earth's crust, 15:1 in humus soil [soil with organic matter], 1:1 in plankton, 1:100 in ferns, and 1:5,000 in mammals."

p. 41, " 'Bardeen was the brains of this joint organism and Brattain was the hands' ": The quote about Bardeen and Brattain being a joint organism comes from the PBS documentary *Transistorized!*

p. 42, "a 'genius sperm bank' ": Shockley's "genius sperm bank," based in California, was officially called the Repository for Germinal Choice. He's the only Nobel Prize winner to admit publicly that he donated, although the sperm bank's founder, Robert K. Graham, claimed a number of others did, too.

p. 46, "Nobel Prize for his integrated circuit": For information on Kilby and the tyranny of numbers, see the wonderful book *The Chip: How Two Americans Invented the Microchip and Launched a Revolution* by T. R. Reid.

Oddly, a club DJ using the handle "Jack Kilby" released a CD in 2006 called *Microchip EP*, with a picture of a very old Kilby on the cover. It features the songs "Neutronium," "Byte My Scarf," "Integrated Circuit," and "Transistor."

3. The Galápagos of the Periodic Table

p. 51, "the reality of atoms": It might seem incredible to us today that Mendeleev refused to believe in atoms, but this was a not uncommon view among chemists at the time. They refused to believe in anything they couldn't see with their own eyes, and they treated atoms as abstractions— a handy way of doing the accounting, maybe, but surely fictitious.

p. 51, "at least in history's judgment?": The best description of the six scientists competing to form the first systematic arrangement of elements can be found in Eric Scerri's *The Periodic Table*. Three other people are generally given credit for coinventing, or at least contributing to, the periodic system.

Alexandre-Emile Béguyer de Chancourtois, according to Scerri, discovered "the single most important step" in developing the periodic table— "that the properties of the elements are a periodic function of their atomic weights, a full seven years before Mendeleev arrived at the same conclusion."

De Chancourtois, a geologist, drew his periodic system on a spiral cylinder, like the thread of a screw. The possibility of his getting credit for the table was dashed when a publisher couldn't figure out how to reproduce the crucial screw diagram showing all the elements. The publisher finally threw his hands up and printed the paper without it. Imagine trying to learn about the periodic table without being able to see it! Nonetheless, de Chancourtois's cause as founder of the periodic system was taken up by his fellow Frenchman Lecoq de Boisbaudran, perhaps partly to get Mendeleev's goat.

William Odling, an accomplished English chemist, seems to have been a victim of bad luck. He got many things right about the periodic table but is virtually forgotten today. Perhaps with his many other chemical and administrative interests, he simply got outworked by Mendeleev, who obsessed over the table. One thing Odling got wrong was the length of the periods of elements (the number of elements that have to appear before similar traits reappear). He assumed all the periods were of length eight, but that's true only at the top of the table. Because of d-shells, rows three and four require a period of eighteen elements. Because of f-shells, rows five and six require thirty-two.

Gustavus Hinrichs was the only American on the list of codiscoverers (although he was not native-born) and the only one described as both a crank and a maverick genius ahead of his time. He published over three thousand scientific articles in four languages and pioneered the study and classification of elements with the light emissions that Bunsen discovered. He also played with numerology and developed a spiral-arm periodic table that placed many really tough elements in the correct groups. As Scerri sums him up, "The work of Hinrichs is so idiosyncratic and labyrinthine that a more complete study will be required before anyone can venture to pronounce on its real value."

p. 54, "Earl Grey 'eats' their utensils": If you're dying to see the gallium practical joke in action, you can see a spoon of gallium melting into nothing on YouTube. Oliver Sacks also talks about pulling pranks of this sort in *Uncle Tungsten*, a memoir of his boyhood.

p. 60, "Streets are named for minerals and elements": For some of the descriptions of the history and geology of Ytterby and for details about the town today, I consulted Jim Marshall, a chemist and historian at the University of North Texas, who was extremely generous with his time and help. He also sent me wonderful pictures. Jim is currently on a quest to revisit the spot where every element was first discovered, which is why he traveled to Ytterby (easy pickings). Good luck, Jim!

4. Where Atoms Come From

p. 65, "proved by 1939": One man who helped figure out the fusion cycles in stars, Hans Bethe, won a $500 prize for doing so, which he used to bribe Nazi officials and spring his mother and, oddly, her furniture from Germany.

p. 66, " 'chemically peculiar stars' ": A fun factoid: Astronomers have identified a strange class of stars that manufacture promethium through an unknown process. The most famous is called Przybylski's star. The truly odd thing is that unlike most fusion events deep inside stars, the promethium must be created on the star's surface. Otherwise, it's too radioactive and short-lived to survive the million-year crawl from the fusion-rich core of a star to its outer layers.

p. 66, "stars govern the fate of mankind": The two portentous Shakespeare quotes that opened the B^2FH paper were as follows:

It is the stars, / The stars above us, govern our conditions.
King Lear, *act 4, scene 3*

The fault, dear Brutus, is not in our stars, / But in ourselves.
Julius Caesar, *act 1, scene 2*

p. 67, "post-ferric fusion": To be technical, stars don't form iron directly. They first form nickel, element twenty-eight, by fusing two atoms of silicon, element fourteen, together. This nickel is unstable, however, and the vast majority of it decays to iron within a few months.

p. 70, "low-watt, brownish light": Jupiter could ignite fusion with deuterium—"heavy" hydrogen with one proton and one neutron—if it had thirteen times its current mass. Given the rarity of deuterium (1 out of every 6,500 hydrogen molecules), it would be a pretty weak star, but it would still count. To ignite regular hydrogen fusion, Jupiter would need seventy-five times its current mass.

p. 72, "like microscopic cubes": And not to be outdone by Jupiter's or Mercury's strange weather, Mars sometimes experiences hydrogen peroxide "snow."

p. 75, "a siderophile, or iron-loving element": The siderophiles osmium and rhenium have also helped scientists reconstruct how the moon was formed from a cataclysmic impact between the very early earth and an asteroid or comet. The moon coalesced from the debris that was thrown up.

p. 78, "later dubbed Nemesis": The goddess Nemesis punished hubris. She made sure no earthly creature could ever grow too proud by striking down anyone who threatened to grow more powerful than the gods. The analogy to the sun's companion star was that if earthly creatures (say, dinosaurs) evolved toward true intelligence, Nemesis would wipe them out before they got traction.

p. 79, "like a carousel as it drifts": Ironically, the overall motion of the sun, if viewed from afar, would resemble the old wheels-within-wheels cycles and epicycles that ancient astronomers bent backward trying to explain in their pre-Copernican, earth-centered cosmos (it's just that earth cannot be called the center anymore, not by a long shot). Like Miescher and proteins, this is an example of the cyclical nature of all ideas, even in science.

5. Elements in Times of War

p. 81, "went on to win the war": For more details on the history of chemical warfare, especially the experience of American troops, see "Chemical Warfare in World War I: The American Experience, 1917–1918," by Major Charles E. Heller, part of the *Leavenworth Papers* published by the Combat Studies Institute, U.S. Army Command and General Staff College, Fort Leavenworth, Kansas, http://www-cgsc.army.mil/carl/resources/csi/Heller/HELLER.asp.

p. 83, "6.7 billion people today": Among the many other things we can attribute to Fritz Haber's ammonia: Charles Townes built the first working maser, the precursor of the laser, by using ammonia as the stimulating agent.

6. Completing the Table ... with a Bang

p. 101, "a full and correct list": Urbain wasn't the only person Moseley embarrassed. Moseley's apparatus also dismantled Masataka Ogawa's claim for discovering nipponium, element forty-three (see chapter 8).

p. 101, "'most irreparable crimes in history'": For accounts of the bungling orders and battles that led to Moseley's death, see *The Making of the Atomic Bomb* by Richard Rhodes. And actually, you should probably just read the whole thing, since it's the best account of twentieth-century science history yet written.

p. 102, "as 'not good for much'": The *Time* magazine article that mentioned the discovery of element sixty-one also included this tidbit about the question of what to name the element: "One convention wag suggested [naming it] grovesium, after loud-mouthed Major General Leslie R. Groves, military chief of the atom bomb project. Chemical symbol: Grr."

p. 104, "Pac-Man style": Besides the electron-gobbling Pac-Man model of the nucleus, scientists at the time also developed the "plum pudding" model, in which electrons were embedded like raisins in a "pudding" of positive charge (Rutherford disproved this by proving that a compact nucleus existed). After the discovery of fission, scientists discovered the liquid drop model, in which large nuclei split like a drop of water on a surface splitting cleanly into two drops. Lise Meitner's work was crucial in developing the liquid drop model.

p. 108, "'would go thermonuclear'": The quotes from George Dyson can be found in his book *Project Orion: The True Story of the Atomic Spaceship*.

p. 109, "'methodological map'": The quote about the Monte Carlo method being a "netherland at once nowhere and everywhere on the usual methodological map" appears in Peter Louis Galison's *Image and Logic*.

7. *Extending the Table, Expanding the Cold War*

p. 115, "'Talk of the Town' section": The *New Yorker* item appeared in the April 8, 1950, issue and was written by E. J. Kahn Jr.

p. 121, "the alarm one last time": For more details about the experiments that led to elements 94 through 110, and for personal information about the man himself, see Glenn Seaborg's autobiographies, especially *Adventures in the Atomic Age* (cowritten with his son Eric). The book is intrinsically interesting because Seaborg was at the center of so much important science and played such a large role in politics for decades. Honestly, though, Seaborg's cautious writing style makes the book a bit bland at points.

p. 124, "poisonous nickel smelters": The information about the lack of trees around Norilsk comes from Time.com, which in 2007 named Norilsk one of the ten most polluted cities in the world. See http://www.time.com/time/specials/2007/article/0,28804,1661031_1661028_1661022,00.html.

p. 130, "June 2009, copernicium (Cn)": It covers a bit of the same material as here, but a story I wrote for Slate.com in June 2009 ("Periodic Discussions," http://www.slate.com/id/2220300/) examines in detail why it took thirteen full years to promote copernicium from provisional element to full member of the periodic table.

8. *From Physics to Biology*

p. 136, "they won forty-two": Besides Segrè, Shockley, and Pauling, the other twelve scientists on the cover of *Time* were George Beadle, Charles Draper, John Enders, Donald Glaser, Joshua Lederberg, Willard Libby, Edward Purcell, Isidor Rabi, Edward Teller, Charles Townes, James Van Allen, and Robert Woodward.

The *Time* "Men of the Year" article contained the following words by Shockley on race. He meant them as complimentary, obviously, but his view on Bunche had to have sounded weird even at the time, and in retrospect it's creepy. "William Shockley, 50, is that rare breed of scientist, a theorist who makes no apology for a consuming interest in the practical applications of his work. 'Asking how much of a research job is pure and how much applied,' says Shockley, 'is like asking how much Negro and white blood Ralph Bunche might have. What's important is that Ralph Bunche is a great man.'"

The article also shows that the legend about Shockley as the main inventor of the transistor was already firmly established:

Hired by Bell Telephone Laboratories right after he graduated from M.I.T. in 1936, theoretical physicist Shockley was one of a team that found a use for what had previously been a scientific parlor stunt: the use of silicon and germanium as a photoelectric device. Along with his partners, Shockley won a Nobel Prize for turning hunks of germanium into the first transistors, the educated little crystals that are fast replacing vacuum tubes in the country's booming electronics industry.

p. 141, "of all the damned luck, Ida Noddack": Overall, Ida Noddack had a spotty run as a chemist. She helped find element seventy-five, but her group's work with element forty-three was riddled with mistakes. She predicted nuclear fission years before anyone else, but about that same time, she began arguing that the periodic table was a useless relic, because the multiplication of new isotopes was rendering it unwieldy. It's not clear why Noddack believed that each isotope was its own element, but she did, and she tried to convince others that they should scrap the periodic system.

p. 142, "'The reason for our blindness is not clear'": The quote from Segrè about Noddack and fission comes from his biography *Enrico Fermi: Physicist*.

p. 144, "a malfunctioning molecule": Pauling (with colleagues Harvey Itano, S. Jonathan Singer, and Ibert Wells) determined that defective hemoglobin causes sickle-cell anemia by running defective cells through a gel in an electric field. Cells with healthy hemoglobin traveled one way in the electric field, while sickle cells moved in the opposite direction. This meant that the two types of molecules had opposite electric charges, a difference that could arise only on a molecular, atom-by-atom level.

Funnily enough, Francis Crick later cited the paper in which Pauling laid out his theory about the molecular basis of sickle-cell anemia as a major influence on him, since it was exactly the sort of nitty-gritty molecular biology that interested Crick.

p. 145, "a molecular appendix": Interestingly, biologists are slowly coming back around to their original view from Miescher's day that proteins are the be-all and end-all of genetic biology. Genes occupied scientists for decades, and they'll never really go away. But scientists now realize that genes cannot account for the amazing complexity of living beings and that far more is going on. Genomics was important fundamental work, but proteomics is where there's real money to be made.

p. 146, "DNA was": To be scrupulous, the 1952 virus experiments with sulfur and phosphorus, conducted by Alfred Hershey and Martha Chase, were not the first to prove that DNA carries genetic information. That honor goes to work with bacteria done by Oswald Avery, published in 1944. Although Avery illuminated the true role of DNA, his work was not widely believed at first. People were beginning to accept it by 1952, but only after the Hershey-Chase experiments did people such as Linus Pauling really get involved in DNA work.

People often cite Avery—and Rosalind Franklin, who unwittingly told Watson and Crick that DNA was a double helix—as prime examples of people who got locked out of Nobel Prizes. That's not quite accurate. Those two scientists never won, but both had died by 1958, and no one won a Nobel Prize for DNA until 1962. Had they still been alive, at least one of them might have shared in the spoils.

p. 147, "James Watson and Francis Crick": For primary documents related to Pauling and his competition with Watson and Crick, see the wonderful site set up by Oregon State University, which has archived and posted the contents of hundreds of personal papers and letters by Pauling and also produced a documentary history called "Linus Pauling and the Race for DNA" at http://osulibrary.oregonstate.edu/specialcollections/coll/pauling/dna/index.html.

p. 149, "before Pauling recovered": After the DNA debacle, Ava Pauling, Linus's wife, famously scolded him. Assuming that he would decipher DNA, Pauling had not broken much of a sweat on his calculations at first, and Ava lit into him: "If [DNA] was such an important problem, why didn't you work harder at it?" Even so, Linus loved her deeply, and perhaps one reason he stayed at Cal Tech so long and never transferred his allegiance to Berkeley, even though the latter was a much stronger school at the time, was that one of the more prominent members of the Berkeley faculty, Robert Oppenheimer, later head of the Manhattan Project, had tried to seduce Ava, which made Linus furious.

p. 150, "the Nobel Prize in Physics": As one last punch in the gut, even Segrè's Nobel Prize was later tainted by accusations (possibly unfounded) that he stole ideas while designing the experiments to discover the antiproton. Segrè and his colleague, Owen Chamberlain, acknowledged working with the combative physicist Oreste Piccioni on methods to focus and guide particle beams with magnets, but they denied that Piccioni's ideas were of much use, and they didn't list him as an author on a crucial paper. Piccioni later helped discover the antineutron. After Segrè and Chamberlain won the prize in 1959, Piccioni remained bitter about the slight for years and finally filed a $125,000 lawsuit against them in 1972— which a judge threw out not for lack of scientific standing but because it had been filed more than a decade after the fact.

From the *New York Times* obituary of Piccioni on April 27, 2002: "'He'd break down your front door and tell you he's got the best idea in the world,' said Dr. William A. Wenzel, a senior scientist emeritus at Lawrence Berkeley National Laboratory who also worked on the antineutron experiment.

'Knowing Oreste, he has a lot of ideas; he throws them out a dozen a minute. Some of them are good, some of them aren't. Nevertheless, I felt he was a good physicist and he contributed to our experiment.'"

9. Poisoner's Corridor

p. 157, "a gruesome record": People still die of thallium poisoning today. In 1994, Russian soldiers working at an old cold war weapons depot found a canister of white powder laced with this element. Despite not knowing what it was, they powdered their feet with it and blended it with their tobacco. A few soldiers reportedly even snorted it. All of them came down with a mysterious, entirely unforeseeable illness, and a few died. On a sadder note, two children of Iraqi fighter pilots died in early 2008 after eating a birthday cake laced with thallium. The motive for the poisoning was unclear, although Saddam Hussein had used thallium during his dictatorship.

p. 161, "in his mother's backyard": Various newspapers in Detroit have tracked David Hahn over the years, but for the most detailed account of Hahn's story, see Ken Silverstein's article in *Harper's* magazine, "The Radioactive Boy Scout" (November 1998). Silverstein later expanded the article into a book of the same name.

10. Take Two Elements, Call Me in the Morning

p. 168, "a cheaper, lighter copper nose": In addition to studying the crust around Brahe's fake nose, the archaeologists who dug him up also found signs of mercury poisoning in his mustache—probably a result of his active research into alchemy. The usual story of Brahe's demise is that he died of a ruptured bladder. One night at a dinner party with some minor royalty, Brahe drank too much, but he refused to get up and go to the bathroom because he thought leaving the table before his social superiors did would be rude. By the time he got home, hours later, he couldn't pee anymore, and he died eleven excruciating days later. The story has become a legend, but it's possible that mercury poisoning contributed as much or more to the astronomer's death.

p. 169, "are copper-coated": The elemental compositions of U.S. coins: New pennies (since 1982) are 97.5 percent zinc but have a thin copper coating, to sterilize the part you touch. (Old pennies were 95 percent copper.) Nickels are 75 percent copper, the balance nickel. Dimes, quarters, and half-dollars are 91.67 percent copper, the balance nickel. Dollar coins (besides special-issue gold coins) are 88.5 percent copper, 6 percent zinc, 3.5 percent manganese, and 2 percent nickel.

p. 169, "one-oared rowboats": Some further facts about vanadium: Some creatures (no one knows why) use vanadium in their blood instead of iron, which turns their blood red or apple green. It can also turn human tongues green. When sprinkled into steel, vanadium greatly strengthens the alloy without adding much weight (much like molybdenum and tungsten; see chapter 5). In fact, Henry Ford once boomed: "Why, without vanadium there would be no automobiles!"

p. 170, "forced to double up": The bus metaphor for how electrons fill their shells one at a time until "someone" is absolutely forced to double up is one of the best in chemistry, both folksy and accurate. It originated with Wolfgang Pauli, who discovered the Pauli "exclusion principle" in 1925.

p. 171, "surgical strikes without surgery": Besides gadolinium, gold is often cited as the best hope for treating cancer. Gold absorbs infrared light that otherwise passes through the body, and grows extremely warm as it does so. Delivering gold-coated particles into tumor cells could allow doctors to fry the tumors without damaging the surrounding tissue. This method was invented by John Kanzius, a businessman and radio technician who underwent thirty-six rounds of chemotherapy for leukemia beginning in 2003. He felt so nauseated and beaten up by the chemo—and was so filled with despair at the sight of the children with cancer he encountered in his hospital—that he decided there had to be a better way. In the middle of the night, he came up with the idea of heating metal particles, and he built a prototype machine using his wife's baking pans. He tested it by injecting half of a hot dog with a solution of dissolved metals and placing it in a chamber of intense radio waves. The tampered-with side of the hot dog fried, while the other half remained cold.

p. 171, "selling it as a supplement": In the May 2009 issue of *Smithsonian*, the article "Honorable Mentions: Near Misses in the Genius Department" describes one Stan Lindberg, a daringly experimental chemist who took it upon himself "to consume every single element of the periodic table." The article notes, "In addition to holding the North American record for mercury poisoning, his gonzo account of a three-week ytterbium bender...('Fear and Loathing in the Lanthanides') has become a minor classic."

I spent a half hour hungrily trying to track down "Fear and Loathing in the Lanthanides" before realizing I'd been had. The piece is pure fiction. (Although who knows? Elements are strange creatures, and ytterbium might very well get you high.)

p. 172, "self-administer 'drugs' such as silver once more": *Wired* magazine ran a short news story in 2003 about the online reemergence of "silver health scams." The money quote: "Meanwhile, doctors across the country have seen a surge in argyria cases. 'In the last year and a half, I've seen six cases of silver poisoning from these so-called health supplements,' said Bill Robertson, the medical director of the Seattle Poison Center. 'They were the first cases I'd seen in fifty years of medical practice.'"

p. 175, "only one handedness, or 'chirality'": It's a bit of a stretcher to claim that people are exclusively left-handed on a molecular level. Even though all of our proteins are indeed left-handed, all of our carbohydrates, as well as our DNA, have a right-handed twist. Regardless, Pasteur's main point remains: in different contexts, our bodies expect and can only process molecules of a specific handedness. Our cells would not be able to translate left-handed DNA, and if we were fed left-handed sugars, our bodies would starve.

p. 177, "the boy lived": Joseph Meister, the little boy Pasteur saved from rabies, ended up becoming the groundskeeper for the Pasteur Institute. Tragically, poignantly, he was still groundskeeper in 1940 when German soldiers overran France. When one officer demanded that Meister, the man with the keys, open up Pasteur's crypt so that he, the officer, could view Pasteur's bones, Meister committed suicide rather than be complicit in this act.

p. 180, "by I. G. Farbenindustrie": The company Domagk worked for, I. G. Farbenindustrie (IGF), would later become notorious around the world for manufacturing the insecticide Zyklon B, which the Nazis used to gas concentration camp prisoners (see chapter 5). The company was broken up shortly after World War II, and many of its directors faced war crimes charges at Nuremberg (*United States v. Carl Krauch, et al.*) for enabling the Nazi government in its aggressive war and mistreating prisoners and captured soldiers. IGF's descendants today include Bayer and BASF.

p. 181, "'the chemistry of dead matter and the chemistry of living matter'": Nevertheless, the universe seems to be chiral on other levels, too, from the subatomic to the supergalactic. The radioactive beta decay of cobalt-60 is an asymmetric process, and cosmologists have seen preliminary evidence that galaxies tend to rotate in counterclockwise spiral arms above our northern galactic pole and in clockwise spirals beneath Antarctica.

p. 182, "the most notorious pharmaceutical of the twentieth century": A few scientists recently reconstructed why thalidomide's devastating effects slipped through clinical trials. For nitty-gritty molecular reasons, thalidomide doesn't cause birth defects in litters of mice, and the German company that produced thalidomide, Grünenthal, did not follow up mouse trials with careful human trials. The drug was never approved for pregnant women in the United States because the head of the Food and Drug Administration, Frances Oldham Kelsey, refused to bow to lobbying pressure to push it through. In one of those curious twists of history, thalidomide is now making a comeback to treat diseases such as leprosy, where it's remarkably effective. It's also a good anticancer agent because it limits the growth of tumors by preventing new blood vessels from forming—which is also why it caused such awful birth defects, since embryos' limbs couldn't get the nutrients they needed to grow. Thalidomide still has a long road back to respectability. Most governments have strict protocols in place to make sure doctors do not give the drug to women of childbearing age, on the off chance that they might become pregnant.

p. 183, "don't know to make one hand or the other": William Knowles unfolded the molecule by breaking a double bond. When carbon forms double bonds, it has only three "arms" coming out of it: two single bonds and a double. (There are still eight electrons, but they are shared over three bonds.) Carbon atoms with double bonds usually form triangular molecules, since a tricornered arrangement keeps its electrons as far apart as possible (120 degrees). When the double bond breaks, carbon's three arms become four. In that case, the way to keep electrons as far apart as possible is not with a planar square but with a three-dimensional tetrahedron. (The vertices in a square are 90 degrees apart. In a tetrahedron, they're 109.5 degrees apart.) But the extra arm can sprout above or below the molecule, which will in turn give the molecule either left- or right-handedness.

11. How Elements Deceive

p. 188, "in underground particle accelerators": A professor of mine from college once held me captive with a story about how a few people died from nitrogen asphyxiation in a particle accelerator at Los Alamos in the 1960s, under circumstances very similar to the NASA accident. After the deaths at Los Alamos, my professor added 5 percent carbon dioxide to the gaseous mixtures in the accelerators he worked on, as a safety measure. He later wrote to me, "Incidentally I did put it to the test about a year later, when one of our graduate student operators did exactly the same thing [i.e., forgot to pump the inert air out and let oxygenated air back in]. I entered the pressure vessel with it full of inert gas.... But not really, [because] by the time I got my shoulders through the hole I was already in desperation, panting due to 'breathe more!' commands from my breathing center." Air is normally 0.03 percent CO_2, so one breath of the doped air was about 167 times more potent.

p. 192, "scales up very quickly to toxic": To its shame and embarrassment, the U.S. government admitted in 1999 that it had knowingly exposed up to twenty-six thousand scientists and technicians to high levels of powdered beryllium, to the extent that hundreds of them developed chronic beryllium disease and related ailments. Most of the people poisoned worked in aerospace, defense, or atomic energy—industries the government decided were too important to arrest or impede, so it neither improved safety standards nor developed an alternative to beryllium. The *Pittsburgh Post-Gazette* ran a long and damning front-page exposé on Tuesday, March 30, 1999. It was titled "Decades of Risk," but one of the subtitles captures the pith of the story better: "Deadly Alliance: How Industry and Government Chose Weapons over Workers."

p. 194, "and calcium": However, scientists at the Monell Chemical Senses Center in Philadelphia believe that in addition to sweet, sour, salty, bitter, and savory (umami), humans have a separate, unique taste for calcium, too. They've definitely found it in mice, and some humans respond to calcium-enriched water as well. So what does calcium taste like? From an announcement about the findings: " 'Calcium tastes calciumy,' [lead scientist Michael] Tordoff said. 'There isn't a better word for it. It is bitter, perhaps even a little sour. But it's much *more* because there are actual receptors for calcium.' "

p. 195, "like so much sand": Sour taste buds can also go flat. These taste buds respond mostly to the hydrogen ion, H^+, but in 2009 scientists discovered that they can taste carbon dioxide as well. (CO_2 combines with H_2O to make a weak acid, H_2CO_3, so perhaps that's why these taste buds perk up.) Doctors discovered this because some prescription drugs, as a side effect, suppress the ability to taste carbon dioxide. The resulting medical condition is known as the "champagne blues," since all carbonated beverages taste flat.

12. Political Elements

p. 206, "killed Pierre": Pierre might not have lived long anyway. In a poignant memory, Rutherford once recalled watching Pierre Curie do an astounding glow-in-the-dark experiment with radium. But in the feeble green glow, the alert Rutherford noticed scars covering Pierre's swollen, inflamed fingers and saw how difficult it was for him to grasp and manipulate a test tube.

p. 207, "her rocky personal life": For more details about the Curies especially, see Sheilla Jones's wonderful book *The Quantum Ten*, an account of the surprisingly contentious and fractious early days of quantum mechanics, circa 1925.

p. 207, "pre-seeped bottles of radium and thorium water": The most famous casualty of the radium craze was the steel tycoon Eben Byers, who drank a bottle of Radithor's radium water every day for four years, convinced it would provide him with something like immortality. He ended up wasting away and dying from cancer. Byers wasn't any more fanatical about radioactivity than a lot of people; he simply had the means to drink as much of the water as he wished. The *Wall Street Journal* commemorated his death with the headline, "The Radium Water Worked Fine Until His Jaw Came Off."

p. 213, "its spot on the table": For the true story of hafnium's discovery, see Eric Scerri's *The Periodic Table*, a thorough and superbly documented account of the rise of the periodic system, including the often strange philosophies and worldviews of the people who founded it.

p. 214, "special 'heavy' water": Hevesy performed heavy-water experiments on goldfish as well as himself, and he ended up killing a number of them.

Gilbert Lewis also used heavy water in a last-ditch effort to win the Nobel Prize in the early 1930s. Lewis knew that Harold Urey's discovery of deuterium—heavy hydrogen with an extra neutron—would win the Nobel Prize, as did every other scientist in the world, including Urey. (After a mostly lackluster career that included ridicule from his in-laws, he came home right after discovering deuterium and told his wife, "Honey, our troubles are over.")

Lewis decided to hitch himself to this no-miss prize by investigating the biological effects of water made with heavy hydrogen. Others had the same idea, but Berkeley's physics department, headed by Ernest O. Lawrence, happened to have the world's largest supply of heavy water, quite by accident. The team had a tank of water it had been using for years in radioactivity experiments, and the tank had a relatively high concentration of heavy water (a few ounces). Lewis begged Lawrence to let him purify the heavy water, and Lawrence agreed—on the condition that Lewis give it back after his experiments, since it might prove important in Lawrence's research, too.

Lewis broke his promise. After isolating the heavy water, he decided to give it to a mouse and see what happened. One queer effect of heavy water is that, like ocean water, the more you drink, the more throat-scratchingly thirsty you feel, since the body cannot metabolize it. Hevesy ingested heavy water in trace amounts, so his body really didn't notice, but Lewis's mouse gulped all the heavy water in a few hours and ended up dead. Killing a mouse was hardly a Nobel Prize–worthy exercise, and Lawrence went apoplectic when he learned a lousy rodent had peed away all his precious heavy water.

p. 216, "blocked him for personal reasons": Kazimierz Fajans's son Stefan Fajans, now a professor emeritus of internal medicine at the University of Michigan's medical school, kindly supplied information to me in an e-mail:

> In 1924 I was six years old, but either then and certainly in the years to follow I did hear from my father of some aspects of the Nobel Prize story. That a Stockholm newspaper published a headline "K. Fajans to Receive Nobel Prize" (I do not know whether it was in chemistry or physics) is not rumor but fact. I remember seeing a copy of that newspaper. I also remember seeing in that newspaper a photo of my father walking in front of a building in Stockholm (probably taken earlier) in somewhat formal dress but not [formal] for that time.... What I did hear was that an influential member of the committee

blocked the award to my father for personal reasons. Whether that was rumor or fact is impossible to know unless someone could look at the minutes of these meetings. I believe they are secret. I do know as a fact that my father expected to receive the Nobel Prize as intimated to him by some people in the know. He expected to receive it in the years to follow.... but it never happened, as you know.

p. 216, "'protactinium' stuck": Meitner and Hahn actually named their element "protoactinium," and only in 1949 did scientists shorten it by removing the extra *o*.

p. 220, "'disciplinary bias, political obtuseness, ignorance, and haste'": There's a wonderful dissection of Meitner, Hahn, and the awarding of the Nobel Prize in the September 1997 issue of *Physics Today* ("A Nobel Tale of Postwar Injustice" by Elisabeth Crawford, Ruth Lewin Sime, and Mark Walker). The article is the source of the quote about Meitner losing the prize because of "disciplinary bias, political obtuseness, ignorance and haste."

p. 221, "the peculiar rules for naming elements": Once a name has been proposed for an element, the name gets only one shot at appearing on the periodic table. If the evidence for the element falls apart, or if the international governing body of chemistry (IUPAC) rules against an element's name, it is blacklisted. This might feel satisfying in the case of Otto Hahn, but it also means that no one can ever name an element "joliotium" after Irène or Frédéric Joliot-Curie, since "joliotium" was once an official candidate name for element 105. It's unclear whether "ghiorsium" has another shot. Perhaps "alghiorsium" would work, although IUPAC frowns on using first and last names, and in fact once rejected "nielsbohrium" in favor of plain "bohrium" for element 107—a decision that didn't please the West German team that discovered 107, since "bohrium" sounds too much like boron and barium.

13. Elements as Money

p. 227, "in Colorado in the 1860s": The fact that gold-tellurium compounds were discovered in the mountains of Colorado is reflected in the name of a local mining town, Telluride, Colorado.

p. 231, "It's called fluorescence": To clarify some easily (and often) confused terms, "luminescence" is the umbrella term for a substance absorbing and emitting light. "Fluorescence" is the instantaneous process described in this chapter. "Phosphorescence" is similar to fluorescence—it consists of molecules absorbing high-frequency light and emitting low-frequency light—but phosphorescing molecules absorb light like a battery and continue to glow long after the light shuts off. Obviously, both fluorescence and phosphorescence derive from elements on the periodic table, fluorine and phosphorus, the two most prominent elements in the molecules that first exhibited these traits to chemists.

p. 237, "the silicon semiconductor revolution eighty years later": Moore's law says that the number of silicon transistors on a microchip will double every eighteen months—amazingly, it has held true since the 1960s. Had the law held for aluminium, Alcoa would have been producing 400,000 pounds of aluminium per day within two decades of starting up, not just 88,000. So aluminium did well, but not quite well enough to beat its neighbor on the periodic table.

p. 237, "Alcoa shares worth $30 million": There's some discrepancy about the magnitude of Charles Hall's wealth at his death. Thirty million dollars is the high end of the range. The confusion may be because Hall died in 1914 but his estate was not settled until fourteen years later. One-third of his estate went to Oberlin College.

p. 237, "spelling disagreement": Aside from differences *between* languages, other spelling discrepancies *within* a language occur with cesium, which the British tend to spell "caesium," and sulfur, which many people still spell "sulphur." You could make a case that element 110 should be spelled mendel*ee*vium, not mendelevium, and that element 111 should be spelled röntgenium, not roentgenium.

14. Artistic Elements

p. 238, "Sybille Bedford could write": The Sybille Bedford quote comes from her novel *A Legacy.*

p. 239, "a hobby": Speaking of strange hobbies, I can't *not* share this in a book full of quirky stories about elements. This anagram won the Special Category prize for May 1999 at the Web site Anagrammy.com, and as far as I'm concerned, this "doubly-true anagram" is the word puzzle of the millennium. The first half equates thirty elements on the periodic table with thirty other elements:

hydrogen + zirconium + tin + oxygen + rhenium + platinum + tellurium + terbium + nobelium + chromium + iron + cobalt + carbon + aluminum + ruthenium + silicon + ytterbium + hafnium + sodium + selenium + cerium + manganese + osmium + uranium + nickel + praseodymium + erbium + vanadium + thallium + plutonium
=
nitrogen + zinc + rhodium + helium + argon + neptunium + beryllium + bromine + lutetium + boron + calcium + thorium + niobium + lanthanum + mercury + fluorine + bismuth + actinium + silver + cesium + neodymium + magnesium + xenon + samarium + scandium + europium + berkelium + palladium + antimony + thulium

That's pretty amazing, even if the number of *ium* endings mitigated the difficulty a little. The kicker is that if you replace each element with its atomic number, the anagram still balances.

$$1 + 40 + 50 + 8 + 75 + 78 + 52 + 65 + 102 + 24 + 26 + 27 + 6 + 13 + 44 +$$
$$14 + 70 + 72 + 11 + 34 + 58 + 25 + 76 + 92 + 28 + 59 + 68 + 23 + 81 + 94$$
$$=$$
$$7 + 30 + 45 + 2 + 18 + 93 + 4 + 35 + 71 + 5 + 20 + 90 + 41 + 57 + 80 + 9 +$$
$$83 + 89 + 47 + 55 + 60 + 12 + 54 + 62 + 21 + 63 + 97 + 46 + 51 + 69$$
$$=$$
$$1416$$

As the anagram's author, Mike Keith, said, "This is the longest doubly-true anagram ever constructed (using the chemical elements—or any other set of this type, as far as I know)."

Along these lines, there's also Tom Lehrer's incomparable song "The Elements." He adapted the tune from Gilbert and Sullivan's "I Am the Very Model of a Modern Major-General," and in it he names every element on the periodic table in a brisk eighty-six seconds. Check it out on YouTube: "There's antimony, arsenic, aluminum, selenium..."

p. 241, "'Plutonists'": Plutonists were sometimes called Vulcanists, too, after the fire god Vulcan. This moniker emphasized the role of volcanoes in the formation of rocks.

p. 243, "Döbereiner's pillars": Döbereiner called his groupings of elements not triads but affinities, part of his larger theory of chemical affinities—a term that gave Goethe (who frequently attended Döbereiner's lectures at Jena) the inspiration for the title *Elective Affinities*.

p. 244, "inches close to majesty": Another majestic design inspired by elements is the wooden Periodic Table Table, a coffee table built by Theodore Gray. The table has more than one hundred slots on top, in which Gray has stored samples of every extant element, including many exclusively man-made ones. Of course, he has only minute quantities of some. His samples of francium and astatine, the two rarest natural elements, are actually hunks of uranium. Gray's argument is that somewhere buried deep inside those hunks are at least a few atoms of each one, which is true and honestly about as good as anyone has ever done. Besides, since most of the elements on the table are gray metals, it's hard to tell them apart anyway.

p. 247, "ruthenium began capping every Parker 51 in 1944": For the details about the metallurgy of the Parker 51, see "Who Was That Man?" by Daniel A. Zazove and L. Michael Fultz, which appeared in the fall 2000 issue of *Pennant*, the house publication of the Pen Collectors of America. The article is a wonderful instance of dedicated amateur history—of keeping alive an obscure but charming bit of Americana. Other resources for Parker pen information include Parker51.com and Vintagepens.com.

The famed tip on the Parker 51 was actually 96 percent ruthenium and 4 percent iridium. The company advertised the nibs as being made of super-durable "plathenium," presumably to mislead competitors into thinking that expensive platinum was the key.

p. 248, "which Remington turned around and printed anyway": The text of the letter Twain sent to Remington (which the company printed verbatim) is as follows:

GENTLEMEN: Please do not use my name in any way. Please do not even divulge the fact that I own a machine. I have entirely stopped using the Type-Writer, for the reason that I never could write a letter with it to anybody without receiving a request by return mail that I would not only describe the machine, but state what progress I had made in the use of it, etc., etc. I don't like to write letters, and so I don't want people to know I own this curiosity-breeding little joker.
Yours truly,
Saml. L. Clemens

15. An Element of Madness

p. 255, "pathological science": Credit for the phrase "pathological science" goes to chemist Irving Langmuir, who gave a speech about it in the 1950s. Two interesting notes on Langmuir: He was the younger, brighter colleague whose Nobel Prize and impudence at lunch might have driven Gilbert Lewis to kill himself (see chapter 1). Later in life, Langmuir grew obsessed with controlling the weather by seeding clouds—a muddled process that skirted awfully close to becoming a pathological science itself. Not even the great ones are immune.

In writing this chapter, I departed somewhat from Langmuir's description of pathological science, which was rather narrow and legalistic. Another take on the meaning of pathological science comes from Denis Rousseau, who wrote a top-rate article called "Case Studies in Pathological Science" for *American Scientist* in 1992. However, I'm also departing from Rousseau, mostly to include sciences such as paleontology that aren't as data driven as other, more famous cases of pathological science.

p. 255, "Philip died at sea": Philip Crookes, William's brother, died on a vessel laying some of the first transatlantic cables for telegraph lines.

p. 257, "supernatural forces": William Crookes had a mystical, pantheistic, Spinozistic view of nature, in which everything partakes of "one sole kind of matter." This perhaps explains why he thought he could commune with ghosts and spirits, since he was part of the same material. If you think about it, though, this view is quite odd, since Crookes made a name for himself discovering new elements—which by definition are different forms of matter!

p. 260, "manganese and the megalodon": For more details on the link between the megalodon and manganese, see Ben S. Roesch, who published an article evaluating how unfeasible it is to think that the megalodon survived in *The Cryptozoology Review* (what a word—"cryptozoology"!) in the autumn of 1998 and revisited the topic in 2002.

p. 261, "The pathology started with the manganese": In another strange link between the elements and psychology, Oliver Sacks notes in *Awakenings* that an overdose of manganese can damage the human brain and cause the same sort of Parkinson's disease that he treated in his hospital. It's a rare cause of Parkinson's, to be sure, and doctors don't quite understand why this element targets the brain instead of, like most toxic elements, going after other vital organs.

p. 264, "a dozen African bull elephants": The bull elephant calculation works as follows. According to the San Diego Zoo, the hugest elephant ever recorded weighed approximately 24,000 pounds. Humans and elephants are made of the same basic thing, water, so their densities are the same. To figure out the relative volume if humans had the appetite of palladium, we can therefore just multiply the weight of a 250-pound man by 900 and divide that number (225,000) by the weight of an elephant. That gives 9.4 elephants swallowed. But remember, that was the biggest elephant ever, standing thirteen feet at the shoulders. The weight of a normal bull elephant is closer to 18,000 pounds, which gives about a dozen swallowed.

p. 268, "a better, more concise description of pathological science": David Goodstein's article on cold fusion was titled "Whatever Happened to Cold Fusion?" It appeared in the fall 1994 issue of the *American Scholar*.

16. Chemistry Way, Way Below Zero

p. 280, "proved an easier thing to blame": The theory that tin leprosy doomed Robert Falcon Scott seems to have originated in a *New York Times* article, although the article floated the theory that what failed was the tins themselves (i.e., the containers) in which Scott's team stored food and other supplies. Only later did people start to blame the disintegration of tin solder. There's an incredibly wide variation, too, in what historians claim that he used for solder, including leather seals, pure tin, a tin-lead mixture, and so on.

p. 281, "and go roaming": Plasma is actually the most common form of matter in the universe, since it's the major constituent of stars. You can find plasmas (albeit very cold ones) in the upper reaches of the earth's atmosphere, where cosmic rays from the sun ionize isolated gas molecules. These rays help produce the eerie natural light shows known as the aurora borealis in the far north. Such high-speed collisions also produce antimatter.

p. 281, "blends of two states": Other colloids include jelly, fog, whipped cream, and some types of colored glass. The solid foams mentioned in chapter 17, in which a gas phase is interspersed throughout a solid, are also colloids.

p. 282, "with xenon in 1962": Bartlett performed the crucial experiment on xenon on a Friday, and the preparation took him the entire day. By the time he broke the glass seal and saw the reaction take place, it was after 7:00 p.m. He was so keyed up that he burst into the hallway in his lab building and began yelling for colleagues. All of them had already gone home for the weekend, and he had to celebrate alone.

p. 285, "Schrieffer": In a macabre late-life crisis, one of the BCS trio, Schrieffer, killed two people, paralyzed another, and injured five more in a horrific car accident on a California highway. After nine speeding tickets, the seventy-four-year-old Schrieffer had had his license suspended, but he decided to drive his new Mercedes sports car from San Francisco to Santa Barbara anyway, and had revved his speed well into the triple digits. Despite his speed, he somehow managed to fall asleep at the wheel and slammed into a van at 111 mph. He was going to be sentenced to eight months in a county jail until the victims' families testified, at which point the judge said that Schrieffer "need[ed] a taste of state prison." The Associated Press quoted his erstwhile colleague Leon Cooper muttering in disbelief: "This is not the Bob I worked with.... This is not the Bob that I knew."

p. 288, "almost": Now, to back off my rigid stance a little, there are a few good reasons why many people conflate the uncertainty principle with the idea that measuring something changes what you're trying to measure—the so-called observer effect. Light photons are about the tiniest tools scientists have to probe things, but photons aren't that much smaller than electrons, protons, or other particles. So bouncing photons off them to measure the size or speed of particles is like trying to measure the speed of a dump truck by crashing a Datsun into it. You'll get information, sure, but at the cost of knocking the dump truck off course. And in many seminal quantum physics experiments, observing a particle's spin or speed or position does alter the reality of the experiment in a spooky way. However, while it's fair to say you have to understand the uncertainty principle to understand any change taking place, the cause of the change itself is the observer effect, a distinct phenomenon.

Of course, it seems likely that the real reason people conflate the two is that we as a society need a metaphor for changing something by the act of observing it, and the uncertainty principle fills that need.

p. 291, "than the 'correct' theory": Bose's mistake was statistical. If you wanted to figure the odds of getting one tail and one head on two coin flips, you could determine the correct answer (one-half) by looking at all four possibilities: HH, TT, TH, and HT. Bose basically treated HT and TH as the same outcome and therefore got an answer of one-third.

p. 293, "the 2001 Nobel Prize": The University of Colorado has an excellent Web site dedicated to explaining the Bose-Einstein condensate (BEC), complete with a number of computer animations and interactive tools: http://www.colorado.edu/physics/2000/bec/.

Cornell and Wieman shared their Nobel Prize with Wolfgang Ketterle, a German physicist who also created the BEC not long after Cornell and Wieman and who helped explore its unusual properties.

Unfortunately, Cornell almost lost the chance to enjoy his life as a Nobel Prize winner. A few days before Halloween in 2004, he was hospitalized with the "flu" and an aching shoulder, and he then slipped into a coma. A simple strep infection had metastasized into necrotizing fasciitis, a severe soft tissue infection often referred to as flesh-eating bacteria. Surgeons amputated his left arm and shoulder to halt the infection, but it didn't work. Cornell remained half-alive for three weeks, until doctors finally stabilized him. He has since made a full recovery.

17. Spheres of Splendor

p. 310, "to study blinking bubbles full-time": Putterman wrote about falling in love with sonoluminescence and his professional work on the subject in the February 1995 issue of *Scientific American*, the May 1998 issue of *Physics World*, and the August 1999 issue of *Physics World*.

p. 312, "bubble science had a strong enough foundation": One theoretical breakthrough in bubble research ended up playing an interesting role in the 2008 Olympics in China. In 1993, two physicists at Trinity University in Dublin, Robert Phelan and Denis Weaire, figured out a new solution to the "Kelvin problem": how to create a bubbly foam structure with the least surface area possible. Kelvin had suggested creating a foam of polygonal bubbles, each of which had fourteen sides, but the Irish duo outdid him with a combination of twelve- and fourteen-sided polygons, reducing the surface area by 0.3 percent. For the 2008 Olympics, an architectural firm drew on Phelan and Weaire's work to create the famous "box of bubbles" swimming venue (known as the Water Cube) in Beijing, which hosted Michael Phelps's incredible performance in the pool.

And lest we be accused of positive bias, another active area of research these days is "antibubbles." Instead of being thin spheres of liquid that trap some air (as bubbles are), antibubbles are thin spheres of air that trap some liquid. Naturally, instead of rising, antibubbles sink.

18. Tools of Ridiculous Precision

p. 317, "calibrate the calibrators": The first step in requesting a new calibration for a country's official kilogram is faxing in a form (1) detailing how you will transport your kilogram through airport security and French customs and (2) clarifying whether you want the BIPM to wash it before and after it has done the measurements. Official kilograms are washed in a bath of acetone, the basic ingredient in fingernail polish remover, then patted dry with lint-free cheesecloth. After the initial washing and after each handling, the BIPM team lets the kilogram stabilize for a few days before touching it again. With all the cleaning and measuring cycles, calibration can easily drag on for months.

The United States actually has two platinum-iridium kilograms, K20 and K4, with K20 being the official copy simply because it has been in the United States' possession longer. The United States also has three all-but-official copies made of stainless steel, two of which NIST acquired within the past few years. (Being stainless steel, they are larger than the dense platinum-iridium cylinders.) Their arrival, coupled with the security headache of flying the cylinders around, explains why Zeina Jabbour isn't in any hurry to send K20 over to Paris: comparing it to the recently calibrated steel cylinders is almost as good.

Three times in the past century, the BIPM has summoned all the official national kilograms in the world to Paris for a mass calibration, but there are no plans to do so again in the near future.

p. 319, "those fine adjustments": To be scrupulous, cesium clocks are based on the *hyperfine* splitting of electrons. The fine splitting of electrons is like a difference of a halftone, while the hyperfine splitting is like a difference of a quarter tone or even an eighth tone.

These days, cesium clocks remain the world standard, but rubidium clocks have replaced them in most applications because rubidium clocks are smaller and more mobile. In fact, rubidium clocks are often hauled around the world to compare and coordinate time standards in different parts of the world, much like the International Prototype Kilogram.

p. 322, "numerology": About the same time that Eddington was working on alpha, the great physicist Paul Dirac first popularized the idea of inconstants. On the atomic level, the electrical attraction between protons and electrons dwarfs the attraction of gravity between them. In fact, the ratio is about 10^{40}, an unfathomable 10,000 trillion trillion trillion times larger. Dirac also happened to be looking at how quickly electrons zoom across atoms, and he compared that fraction of a nanosecond with the time it takes beams of light to zoom across the entire universe. Lo and behold, the ratio was 10^{40}.

Predictably, the more Dirac looked for it, the more that ratio popped up: the size of the universe compared to the size of an electron; the mass of the universe compared to the mass of a proton; and so on. (Eddington also once testified that there were approximately 10^40 times 10^40 protons and electrons in the universe—another manifestation.) Overall, Dirac and others became convinced that some unknown law of physics forced those ratios to be the same. The only problem was that some ratios were based on changing numbers, such as the size of the expanding universe. To keep his ratios equal, Dirac hit upon a radical idea—that gravity grew weaker with time. The only plausible way this could happen was if the fundamental gravitational constant, G, had shrunk.

Dirac's ideas fell apart pretty quickly. Among other flaws that scientists pointed out was that the brightness of stars depends heavily on G, and if G had been much higher in the past, the earth would have no oceans, since the overbright sun would have boiled them away. But Dirac's search inspired others. At the height of this research, in the 1950s, one scientist even suggested that all fundamental constants were constantly diminishing—which meant the universe wasn't getting bigger, as commonly thought, but that the earth and human beings were shrinking! Overall, the history of varying constants resembles the history of alchemy: even when there's real science going on, it's hard to sift it from the mysticism. Scientists tend to invoke inconstants to explain away whatever cosmological mysteries happen to trouble a particular era, such as the accelerating universe.

p. 325, "Australian astronomers": For details about the work of the Australian astronomers, see an article that one of them, John Webb, wrote for the April 2003 issue of *Physics World*, "Are the Laws of Nature Changing with Time?" I also interviewed a colleague of Webb's, Mike Murphy, in June 2008.

p. 326, "a fundamental constant changing": In other alpha developments, scientists have long wondered why physicists around the world cannot agree on the nuclear decay rates of certain radioactive atoms. The experiments are straightforward, so there's no reason why different groups should get different answers, yet the discrepancies persist for element after element: silicon, radium, manganese, titanium, cesium, and so on.

In trying to solve this conundrum, scientists in England noted that groups reported different decay rates at different times of the year. The English group then ingeniously suggested that perhaps the fine structure constant varies as the earth revolves around the sun, since the earth is closer to the sun at certain times of the year. There are other possible explanations for why the decay rate would vary periodically, but a varying alpha is one of the more intriguing, and it would be fascinating if alpha really did vary so much even within our own solar system!

p. 326, "from the beginning": Paradoxically, one group really rooting for scientists to find evidence for a variable alpha is Christian fundamentalists. If you look at the underlying mathematics, alpha is defined in terms of the speed of light, among other things. Although it's a little speculative, the odds are that if alpha has changed, the speed of light has changed, too. Now, everyone, including creationists, agrees that light from distant stars provides a record, or at least appears to provide a record, of events from billions of years ago. To explain the blatant contradiction between this record and the time line in Genesis, some creationists argue that God created a universe with light already "on the way" to test believers and force them to choose God or science. (They make similar claims about dinosaur bones.) Less draconian creationists have trouble with that idea, since it paints God as deceptive, even cruel. However, if the speed of light had been billions of times larger in the past, the problem would evaporate. God still could have created the earth six thousand years ago, but our ignorance about light and alpha obscured that truth. Suffice it to say, many of the scientists working on variable constants are horrified that their work is being appropriated like this, but among the very few people practicing what might be called "fundamentalist physics," the study of variable constants is a hot, hot field.

p. 327, "impish": There's a famous picture of Enrico Fermi at a blackboard, with an equation for the definition of alpha, the fine structure constant, appearing behind him. The queer thing about the picture is that Fermi has the equation partly upside down. The actual equation is alpha = $e^2/\hbar c$, where e = the charge of the electron, $\hbar$ = Planck's constant (h) divided by 2π, and c = the speed of light. The equation in the picture reads alpha = $\hbar^2/ec$. It's not clear whether Fermi made an honest mistake or was having a bit of fun with the photographer.

p. 328, "Drake originally calculated": If you want a good look at the Drake Equation, here goes. The number of civilizations in our galaxy that are trying to get in touch with us, N, supposedly equals

$$N = R^* \times f_p \times n_e \times f_l \times f_i \times f_c \times L$$

where R^* is the rate of star formation in our local galaxy; f_p is the fraction of stars that conjure up planets; n_e is the average number of suitable home planets per conjuring star; f_l, f_i, and f_c are, respectively, the fractions of hospitable planets with life, intelligent life, and sociable, eager-to-communicate life; and L is the length of time alien races send signals into space before wiping themselves out.

The original numbers Drake ran were as follows: our galaxy produces ten stars per year ($R^* = 10$); half of those stars produce planets ($f_p = \frac{1}{2}$); each star with planets has two suitable homes ($n_e = 2$, although our own solar system has seven or so—Venus, Mars, Earth, and a few moons of Jupiter and Saturn); one of those planets will develop life ($f_l = 1$); 1 percent of those planets will achieve intelligent life ($f_i = 1/100$); 1 percent of *those* planets will produce post-caveman life capable of beaming signals into space ($f_c = 1/100$);

and they will do so for ten thousand years (L = 10,000). Work all that out, and you get ten civilizations trying to communicate with earth.

Opinions about those values differ, sometimes wildly. Duncan Forgan, an astrophysicist at the University of Edinburgh, recently ran a Monte Carlo simulation of the Drake Equation. He fed in random values for each of the variables, then computed the result a few thousand times to find the most probable value. Whereas Drake figured that there were ten civilizations trying to get in touch with us, Forgan calculated a total of 31,574 civilizations just in our local galaxy. The paper is available at http://arxiv.org/abs/0810.2222.

19. Above (and Beyond) the Periodic Table

p. 334, "one force gains the upper hand, then the other": The third of the four fundamental forces is the weak nuclear force, which governs how atoms undergo beta decay. It's a curious fact that francium struggles because the strong nuclear force and the electromagnetic force wrestle inside it, yet the element arbitrates the struggle by appealing to the weak nuclear force.

The fourth fundamental force is gravity. The strong nuclear force is a hundred times stronger than the electromagnetic force, and the electromagnetic force is a hundred billion times stronger than the weak nuclear force. The weak nuclear force is in turn ten million billion billion times stronger than gravity. (To give you some sense of scale, that's the same number we used to compute the rarity of astatine.) Gravity dominates our everyday lives only because the strong and weak nuclear forces have such short reach and the balance of protons and electrons around us is equal enough to cancel most electromagnetic forces.

p. 337, "un·bi·bium": After decades of scientists having to build superheavy elements laboriously, atom by atom, in 2008 Israeli scientists claimed to have found element 122 by reverting to old-style chemistry. That is, after sifting through a natural sample of thorium, the chemical cousin of 122 on the periodic table, for months on end, a team led by Amnon Marinov claimed to have identified a number of atoms of the extra-heavy element. The crazy part about the enterprise wasn't just the claim that such an old-fashioned method resulted in a new element; it was the claim that element 122 had a half-life of more than 100 million years! That was so crazy, in fact, that many scientists got suspicious. The claim was looking shakier and shakier, but as of late 2009, the Israelis hadn't backed off from their claims.

p. 337, "once-dominant Latin in science": Regarding the decline of Latin, except on the periodic table: for whatever reason, when a West German team bagged element 108 in 1984, they decided to name it hassium, after the Latin name for part of Germany (Hesse), instead of naming it deutschlandium or some such thing.

p. 341, "rectilinear shapes": It's not a new version of the periodic table, but it's certainly a new way to present it. In Oxford, England, periodic table taxicabs and buses are running people around town. They're painted tires to roof with different columns and rows of elements, mostly in pastel hues. The fleet is sponsored by the Oxford Science Park. You can see a picture at http://www.oxfordinspires.org/newsfromImageWorks.htm.

You can also view the periodic table in more than two hundred different languages, including dead languages like Coptic and Egyptian hieroglyphic, at http://www.jergym.hiedu.cz/~canovm/vyhledav/chemici2.html.

BIBLIOGRAPHY

These were far from the only books I consulted during my research, and you can find more information about my sources in the "Notes and Errata" section. These were simply the best books for a general audience, if you want to know more about the periodic table or various elements on it.

Patrick Coffey. *Cathedrals of Science: The Personalities and Rivalries That Made Modern Chemistry*. Oxford University Press, 2008.
John Emsley. *Nature's Building Blocks: An A–Z Guide to the Elements*. Oxford University Press, 2003.
Sheilla Jones. *The Quantum Ten*. Oxford University Press, 2008.
T. R. Reid. *The Chip: How Two Americans Invented the Microchip and Launched a Revolution*. Random House, 2001.
Richard Rhodes. *The Making of the Atomic Bomb*. Simon & Schuster, 1995.
Oliver Sacks. *Awakenings*. Vintage, 1999.
Eric Scerri. *The Periodic Table*. Oxford University Press, 2006.
Glenn Seaborg and Eric Seaborg. *Adventures in the Atomic Age: From Watts to Washington*. Farrar, Straus and Giroux, 2001.
Tom Zoellner. *Uranium*. Viking, 2009.

INDEX

Note: Elements are listed in bold. Italic page numbers refer to illustrations.

THE PERIODIC TABLE

H 1 1.008								
Li 3 6.941	Be 4 9.012							
Na 11 22.990	Mg 12 24.305							
K 19 39.098	Ca 20 40.078	Sc 21 44.956	Ti 22 47.861	V 23 50.941	Cr 24 51.996	Mn 25 54.938	Fe 26 55.845	Co 27 58.993
Rb 37 85.468	Sr 38 87.621	Y 39 88.906	Zr 40 91.224	Nb 41 92.906	Mo 42 95.942	Tc 43 98.906	Ru 44 101.072	Rh 45 102.905
Cs 55 132.905	Ba 56 137.327	57-71	Hf 72 178.492	Ta 73 180.948	W 74 183.841	Re 75 186.207	Os 76 190.233	Ir 77 192.217
Fr 87 223	Ra 88 226	89-103	Rf 104 (267)	Db 105 (268)	Sg 106 (271)	Bh 107 (270)	Hs 108 (277)	Mt 109 (276)

La 57 138.905	Ce 58 140.116	Pr 59 140.908	Nd 60 144.242	Pm 61 145.0	Sm 62 150.362	Eu 63 151.964
Ac 89 227	Th 90 232.038	Pa 91 231.036	U 92 238.029	Np 93 (237)	Pu 94 (244)	Am 95 (243)

					He 2 4.003
B 5 10.812	C 6 12.011	N 7 14.007	O 8 15.999	F 9 18.998	Ne 10 20.180
Al 13 26.982	Si 14 28.086	P 15 30.974	S 16 32.066	Cl 17 35.453	Ar 18 39.948

Ni 28 58.693	Cu 29 63.546	Zn 30 65.384	Ga 31 69.723	Ge 32 72.641	As 33 74.922	Se 34 78.963	Br 35 79.904	Kr 36 83.798
Pd 46 106.421	Ag 47 107.868	Cd 48 112.412	In 49 114.818	Sn 50 118.711	Sb 51 121.760	Te 52 127.603	I 53 126.904	Xe 54 131.294
Pt 78 195.085	Au 79 196.967	Hg 80 200.592	Tl 81 204.383	Pb 82 207.2	Bi 83 208.980	Po 84 209	At 85 210	Rn 86 222
Ds 110 (281)	Rg 111 (280)	Cn 112 (285)	Uut 113 (284)	Uuq 114 (289)	Uup 115 (288)	Uuh 116 (293)	Uus 117 (294)	Uuo 118 (294)

Gd 64 157.253	Tb 65 158.925	Dy 66 162.500	Ho 67 164.930	Er 68 167.259	Tm 69 168.934	Yb 70 173.043	Lu 71 174.967
Cm 96 (247)	Bk 97 (247)	Cf 98 (251)	Es 99 (252)	Fm 100 (257)	Md 101 (258)	No 102 (259)	Lr 103 (262)